INSTITUTE OF GEOLOGICAL SCIENCES

Natural Environment Research Council

Mineral Assessment Report 84

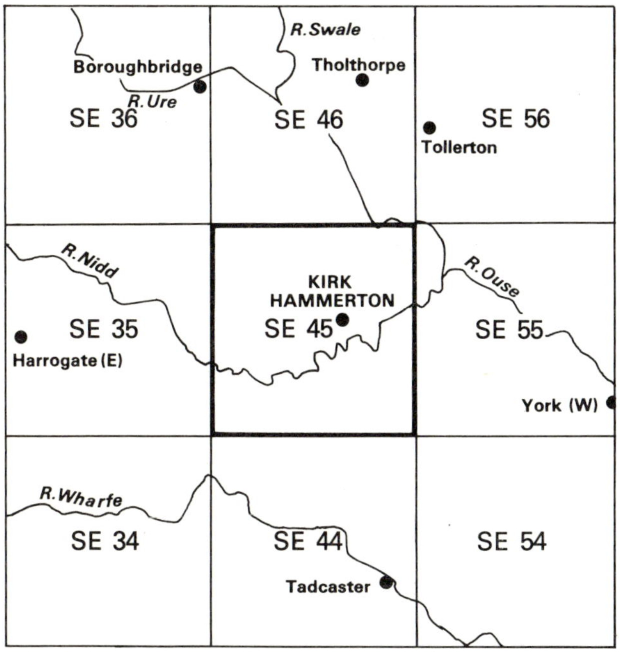

The sand and gravel resources of the country around Kirk Hammerton, North Yorkshire

Description of 1:25 000 resource sheet SE 45

J. R. A. Giles

Contributor
A. H. Cooper

Please ensure ALL parts are returned

ISBN 0 11 884207 2*

London Her Majesty's Stationery Office 1981

The first twelve reports on the assessment of British sand and gravel resources appeared in the Report Series of the Institute of Geological Sciences as a subseries. Report No. 13 and subsequent reports appear as Mineral Assessment Reports of the Institute.

Details of published reports appear at the end of this Report.

Any enquiries concerning this report may be addressed to
Head, Industrial Minerals Assessment Unit,
Institute of Geological Sciences,
Keyworth, Nottingham NG12 5GG.

PREFACE

National resources of many industrial minerals may seem so large that stocktaking appears unnecessary, but the demand for minerals and for land for all purposes is intensifying and it has become increasingly clear in recent years that regional assessments of the resources of these minerals should be undertaken. The publication of information about the quantity and quality of deposits over large areas is intended to provide a comprehensive factual background against which planning decisions can be made.

Sand and gravel, considered together as naturally occurring aggregate, was selected as the bulk mineral demanding the most urgent attention, initially in the south-east of England, where about half the national output is won and very few sources of alternative aggregates are available. Following a short feasability project, initiated in 1966 by the Ministry of Land and Natural Resources, the Industrial Minerals Assessment Unit (formerly the Mineral Assessment Unit) began systematic surveys in 1968. The work is now being financed by the Department of the Environment and is being undertaken with the cooperation of the Sand and Gravel Association of Great Britain.

This report describes the resources of sand and gravel of the country around Kirk Hammerton, North Yorkshire, shown on the accompanying 1:25 000 resource map SE 45. The survey was conducted by Mr. A. M. Harrisson, assisted in the drilling and sampling programme by Mr R. Stanczyszyn; Mr J. R. A. Giles compiled the report. The work is based on a geological survey, at 1:10 000, in 1975-76 by Mr R. Anderton, Dr A. H. Cooper, who contributed the geological account of the district, and Dr G. D. Gaunt. Mr J. W. Gardner, CBE and Mr C. L. Reeves (Land Agents) have been responsible for negotiating access to land for drilling. The ready cooperation of landowners and tenants in this work is gratefully acknowledged.

G. M. Brown
Director

Institute of Geological Sciences
Exhibition Road
London SW7 2DE

1st July 1981

CONTENTS

The sand and gravel resources of the country around Kirk Hammerton, North Yorkshire

Description of 1:25 000 resource sheet SE 45

J. R. A. Giles

SUMMARY

The geological maps of the Institute of Geological Sciences, pre-existing borehole information, and 84 boreholes drilled for the Industrial Minerals Assessment Unit form the basis of the assessment of the sand and gravel resources of the country around Kirk Hammerton, North Yorkshire.

All the deposits in the district that might be potentially workable for sand and gravel have been investigated and a simple statistical method has been used to estimate the volume. The reliability of the volume estimates is given at the symmetrical 95 per cent probability level.

The 1:25 000 map is divided into four resource blocks, containing between 10.6 and 27.9 km^2 of sand and gravel. For each block the geology of the deposits is described, and the mineral-bearing area, the mean thickness of overburden and mineral and the mean gradings are stated. Detailed borehole data are also given. The geology, the position of the boreholes and the outlines of the resource blocks are shown on the accompanying map.

Notes

Each borehole registered with the Institute is identified by a four-element alphanumeric descriptor (e.g. SE 45 NW 16). The first two elements define the 10-km square (of the National Grid) in which the borehole is situated; the third element defines a quadrant of that square, and the fourth is the accession number of the borehole. In the text of the report the borehole is normally referred to by the last two elements alone (e.g. NW 16).

All National Grid references in this publication lie within the 100-km square SE (unless otherwise stated). Grid references are given to eight figures, accurate to within 10 m for borehole locations (In the text, four- and six-figure grid references are used for more extensive locations, for example for farms).

Bibliographical reference

GILES, J. R. A. 1981. The sand and gravel resources of the country around Kirk Hammerton, North Yorkshire: description of 1:25 000 resource sheet SE 45. Miner. Assess. Rep. Inst. Geol. Sci., No. 84.

Author and contributor

J. R. A. Giles, BSc
Institute of Geological Sciences, Keyworth, Nottingham NG12 5GG

A. H. Cooper, PhD
Institute of Geological Sciences, Ring Road Halton, Leeds LS15 8TQ

INTRODUCTION

The survey is concerned with the estimation of resources, which include deposits that are not currently exploitable but have a foreseeable use, rather than reserves, which can only be assessed in the light of current, locally prevailing, economic considerations. Clearly, neither the economic nor the social factors used to decide whether a deposit may be workable in the future can be predicted; they are likely to change with time. Deposits not currently economically workable may be exploited as demand increases, as higher-grade or alternative materials become scarce, or as improved processing techniques are applied to them. The improved knowledge of the main physical properties of the resource and their variability, which this survey seeks to provide, will add significantly to the factual background against which planning policies can be decided (Archer, 1969; Thurrell, 1971, 1981; Harris and others, 1974).

The survey provides information at the 'indicated' level "for which tonnage and grade are computed partly from specific measurements, samples or production data and partly from projection for a reasonable distance on geologic evidence. The sites available for inspection, measurement, and sampling are too widely or otherwise inappropriately spaced to permit the mineral bodies to be outlined completely or the grade established throughout" (Bureau of Mines and Geological Survey, 1948, p. 15).

It follows that the whereabouts of reserves must still be established and their size and quality proved by the customary detailed exploration and evaluation undertaken by the industry. However, the information provided by this survey should assist in the selection of the best targets for such further work. The following arbitrary physical criteria have been adopted:

a The deposit should average at least 1 m in thickness.

b The ratio of overburden to sand and gravel should be no more than 3:1.

c The proportion of fines (particles passing the 0.063 m B.S. sieve) should not exceed 40 per cent.

d The deposit should lie within 25 m of the surface, this being taken as the likely maximum working depth under most circumstances. It follows from the second criterion that boreholes are drilled no deeper than 18 m if no sand and gravel has been proved.

A deposit of sand and gravel that broadly meets these criteria is regarded as 'potentially workable' and is described and assessed as 'mineral' in this report. As the assessment is at the indicated level, parts of such a deposit may not satisfy all the criteria.

Pre-Pleistocene rocks, which are usually consolidated and devoid of potentially workable sand and gravel, are referred to as 'bedrock'; 'waste' is any material other than bedrock or mineral; 'overburden' is waste that occurs between the surface and an underlying body of mineral.

For the particular needs of assessing sand and gravel resources, a grain-size classification based on the geometric scale 1/16 mm, $\frac{1}{4}$ mm, 1 mm, 4 mm, 16 mm, 64 mm has been adopted. The boundaries between fines (that is, the clay and silt fractions) and sand, and between sand and gravel-grade material, are placed at 1/16 mm and 4 mm respectively (see Appendix C).

The volume and other characteristics are assessed within resource blocks, each of which, ideally, contains

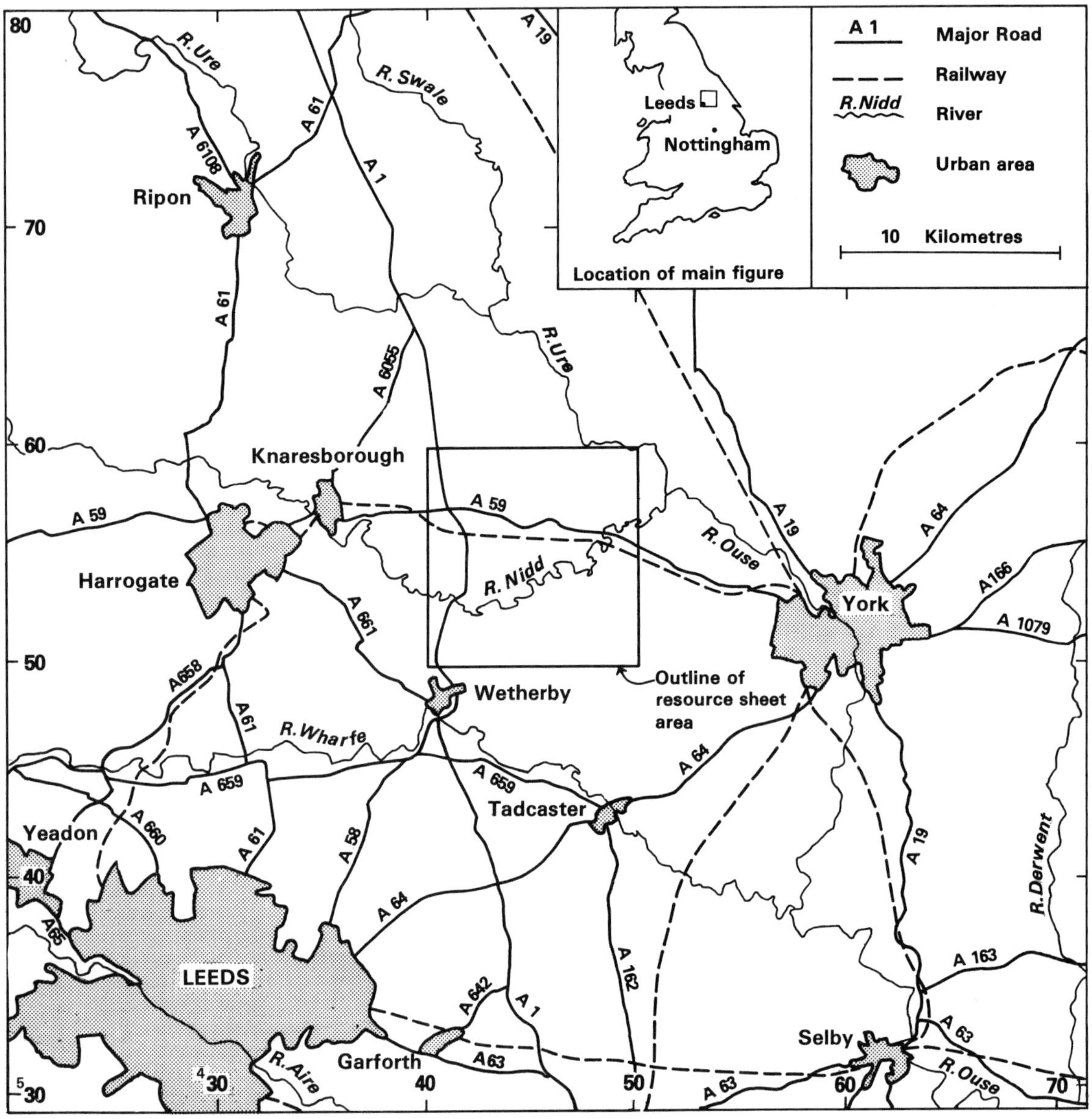

Figure 1 Map showing the location of sheet SE 45.

approximately 10 km² of sand and gravel. No account is taken of any factors, for example roads, villages or land of high agricultural or landscape value, which might stand in the way of sand and gravel being exploited, although towns are excluded. The estimated total volume therefore bears no simple relationship to the amount that could be extracted in practice.

It must be emphasised that the assessment applies to the resource block as a whole; valid conclusions cannot be drawn about mineral in parts of a block, except in the immediate vicinity of the actual sample points.

DESCRIPTION OF THE DISTRICT

TOPOGRAPHY

This report describes the sand and gravel resources of an area of 100 km² in North Yorkshire (Figure 1) on the western margin of the Vale of York between York and Knaresborough. Two major roads serve the district: the A1 trunk road, which runs from north to south near the western border, and the A59 York-Harrogate road, which traverses it from east to west; the York-Harrogate railway runs just south of the A59.

The eastern part of the district, east of Green Hammerton and north-east of Tockwith, is low-lying, with an average elevation of about 15 m above OD. The highest ground is in the north-west, where undulating drift-covered country rises to a maximum elevation of 69 m at Sand Hill [406 597]. South of the Nidd, the ground rises gently to a maximum elevation of 48 m above OD south of Hillside Farm [495 503].

Except for the north-eastern corner, which is drained by small tributaries of the Ouse, the district lies within the drainage basin of the Nidd, which meanders in a generally east-north-easterly direction past Cowthorpe and Kirk Hammerton towards its confluence with the Ouse at Nun Monkton.

The district is almost entirely agricultural, with arable farming predominating; wheat, barley, sugar-beet and potatoes are the main crops. The airfield at Tockwith is no longer used for flying, but a Driver Training Centre has been established there.

GEOLOGY

The Solid rocks and Drift deposits found in the district are listed in Table 1. The relationship between deposits is illustrated in Figure 2. A detailed account of the geology of the southern part of the area is given by Edwards and others (1950).

Table 1 Stratigraphy

DRIFT		
Quaternary	Peat	
	Alluvium	
	River Terrace Deposits, undifferentiated	
	Sand and Gravel	Glacial Lake Deposits
	Silt and Clay	Glacial Lake Deposits
	Fluvio-glacial Terrace Deposits, undifferentiated	
	Glacial Sand and Gravel (and associated laminated clay)	
	Till and Sandy Till	
	Clay associated with Fluvio-glacial and Older River Sand and Gravel	
	Fluvio-glacial and Older River Sand and Gravel	
SOLID		
Triassic	Sherwood Sandstone Group	
Permian	Upper Marl	
	Upper Magnesian Limestone	
	Middle Marl	
	Lower Magnesian Limestone	

SOLID

The Solid rocks are concealed by Drift, except for some small exposures of the Sherwood Sandstone Group between Moor Farm [4438 5970] and Long Marston [4950 5175] and in the bank of the River Nidd at Broad Oak Farm [4423 5276].

Lower Magnesian Limestone This formation incrops against Drift in the south-western corner of the district with a boundary fault that passes beneath Willowgarth Plantation [4310 5042], to the north. It is about 45 m thick and consists of dolomitic limestone.

Middle Marl The Middle Marl incrops to the south of the fault mentioned above. It is about 26 m thick and is composed of reddish brown, silty, calcareous mudstone with interbedded gypsum and anhydrite at depth. These evaporites have commonly undergone solution, causing subsidence at the surface.

Upper Magnesian Limestone This formation forms a faulted dip-slope in the south-west of the area. It is between 12 and 14 m thick and is largely composed of thinly bedded dolomitic limestone.

Upper Marl There are no exposures of the Upper Marl in the district, the entire crop being covered by Drift. Boreholes, however, prove between 19 and 24 m of reddish brown silty mudstone with gypsum and anhydrite at depth. A few thin sandstones are present, mainly near the top of the formation, where it passes into the overlying Sherwood Sandstone.

Sherwood Sandstone Group The lowermost 160 m of the group is present in the district and is composed of reddish brown, fine to coarse-grained sandstones, which are commonly cross-bedded.

DRIFT

Drift deposits cover most of the district and fill a valley system incised to a depth of about 7 m below OD (Figure 3). Fluvio-glacial and Older River Sand and Gravel floor the buried valleys and in the west of the area are associated with clays. These valley deposits are overlain by Devensian glacial deposits which include Till, Sandy Till and Glacial Sand and Gravel. The glacial deposits have an undulating to hummocky topography and in the southern part of the area form part of the York-Escrick Moraines (Edwards and others, 1950). Both to the north and south of these moraines glacial lakes developed when the ice sheet waned, and lacustrine silts, clays, sands and gravels were laid down. After the lakes were drained, the rivers incised their present drainage pattern.

Fluvio-glacial and Older River Sand and Gravel These are up to 10 m thick and partly fill the buried valley system (Figures 2 and 3). The top surface of the deposits falls from about 20 m above OD in the west to 4 m below OD in the north-east. The deposits are generally overlain by Till and the contact is uneven, probably as the result of glacial scour.

Clay associated with Fluvio-glacial and Older River Sand and Gravel Clay, usually laminated and up to 7 m thick, overlies and is locally interdigitated with the sand and gravel in the buried valleys west of Kirk Hammerton and Tockwith. The clay, overlain in most places by Till, is discontinuous and may have been affected by glacial scour.

Till and Sandy Till Till and Sandy Till cover most of the district and overlie both bedrock and the valley-fill deposits. The tills, which may exceed 18 m in thickness, have a matrix of silty clay, sandy clay or 'clayey' sand, and usually vary from reddish brown to dark brown in colour. The Sandy Till is more common in areas underlain by the Sherwood Sandstone Group. Pebbles, cobbles and boulders in the tills are composed dominantly of Carboniferous sandstone, but Carboniferous limestone is locally abundant. Permian limestone fragments are scarce, except in areas where the tills are underlain by Lower and Upper Magnesian Limestones; Sherwood Sandstone fragments are common at depth. Lenses and discontinuous beds of laminated clay, sand and gravel are present within the tills.

Glacial Sand and Gravel These deposits vary considerably in composition, thickness, topographic expression and their relationships to the tills. They range from reddish brown, fine to medium-grained sand with clay laminae to coarse gravel with boulders in a 'clayey' sand matrix (see Appendix C). The gravel is largely composed of Carboniferous sandstone, but Carboniferous limestone is locally dominant. The deposits occur above, within and below the till, and were mostly laid down by water, on top of or within a stagnating ice-sheet. A few 'clayey' gravels, however, are probably of morainic origin.

Small lenses of laminated clay with silt and sand beds occur within the glacial sequence. Thicknesses of up to 3.4 m have been proved for these clays in IMAU boreholes.

Fluvio-glacial Terrace Deposits, undifferentiated In the south-west of the area a terrace of sand and gravel at about 25 m above OD extends from north of Deighton Grange [411 512] to near Deighton Whin [407 500]. This deposit post-dates the tills and is continuous with the fluvio-glacial outwash from the Kirk Deighton Channel about 2 km to the west (Edwards and others, 1950).

Glacial Lake Deposits Clays, silts, sands and gravels of lacustrine origin occur throughout the area, and are especially common in the east. They generally overlie till, occupy low ground, and form flat expanses.

Silt and Clay Stiff, grey to brown, stoneless clay with laminae of silt and fine sand forms the bulk of the lacustrine deposits. In the west of the area these deposits are generally thin, but in the north-east they are proved to be up to 14.7 m thick.

Sand and Gravel Thin sands interbedded with the lacustrine clays are common. Between Walshford [415 532] and Wilstrop Moor [489 535], however, sand and gravel also occur at the surface. These deposits fill channels cut in the clay or form elevated lenses on the

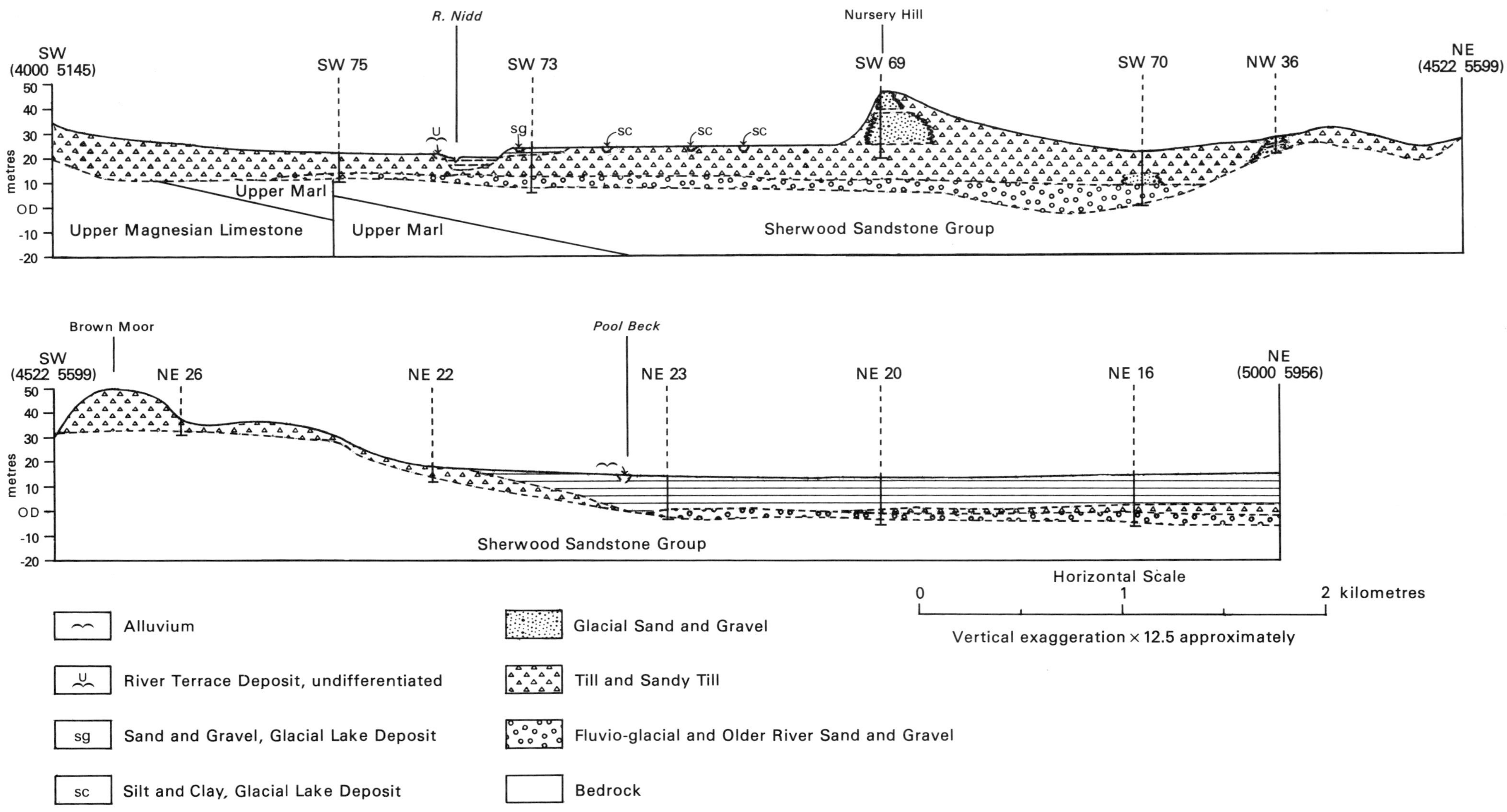

Figure 2 Schematic section across the district. The line of the sections is shown on the resource map.

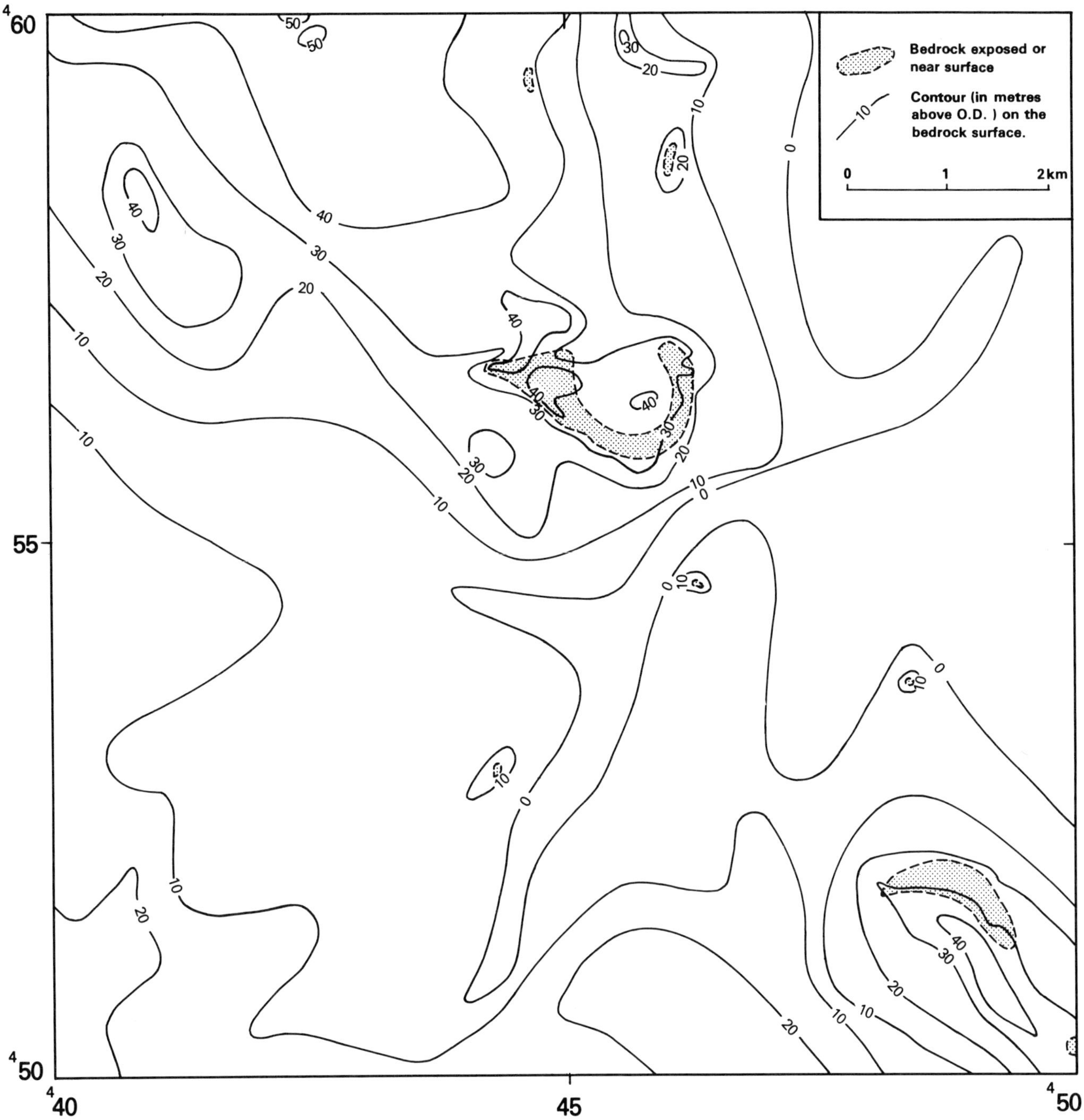

Figure 3 Contour map showing the form of the bedrock surface.

surface of the clay. They are concentrated where the River Nidd enters the lacustrine flat and are probably delta and levee deposits. In the south-west around Deighton Whin [409 501] sand and gravel form a beach facies marginal to the clay flat.

River Terrace Deposits, undifferentiated Several terraces flank the River Nidd west of Kirk Hammerton. The higher terraces are generally composed of 'clayey' sand and gravel, and the lower ones of silt and clay.

Alluvium Clay and silt with local lenses of sand form the river alluvium, which is up to 8 m thick. Sand and gravel are commonly present at the base, laid down as channel lag deposits.

Numerous ill-drained hollows contain alluvial silt, commonly associated with peat. The majority are situated in localities mantled by glacial deposits and some are typical kettle holes. A few, confined to localitites underlain by Middle Marl, Upper Magnesian Limestone or Upper Marl, are probably subsidence features resulting from evaporite solution.

Peat This deposit occupies low-lying, ill-drained land and is generally less than 3 m thick. South of Lingcroft [442 511] it forms several extensive flats and is associated with laminated clay.

COMPOSITION OF THE SAND AND GRAVEL DEPOSITS

Seven deposits contain potentially workable sand and gravel within the district: Fluvio-glacial and Older River Sand and Gravel, Sandy Till, Glacial Sand and Gravel, Fluvio-glacial Terrace Deposits, Glacial Lake Deposits (Sand and Gravel), River Terrace Deposits, and Alluvium.

Fluvio-glacial and Older River Sand and Gravel This deposit varies from sand to gravel which may be clean to 'very clayey'; however, its mean grading of fines 9 per

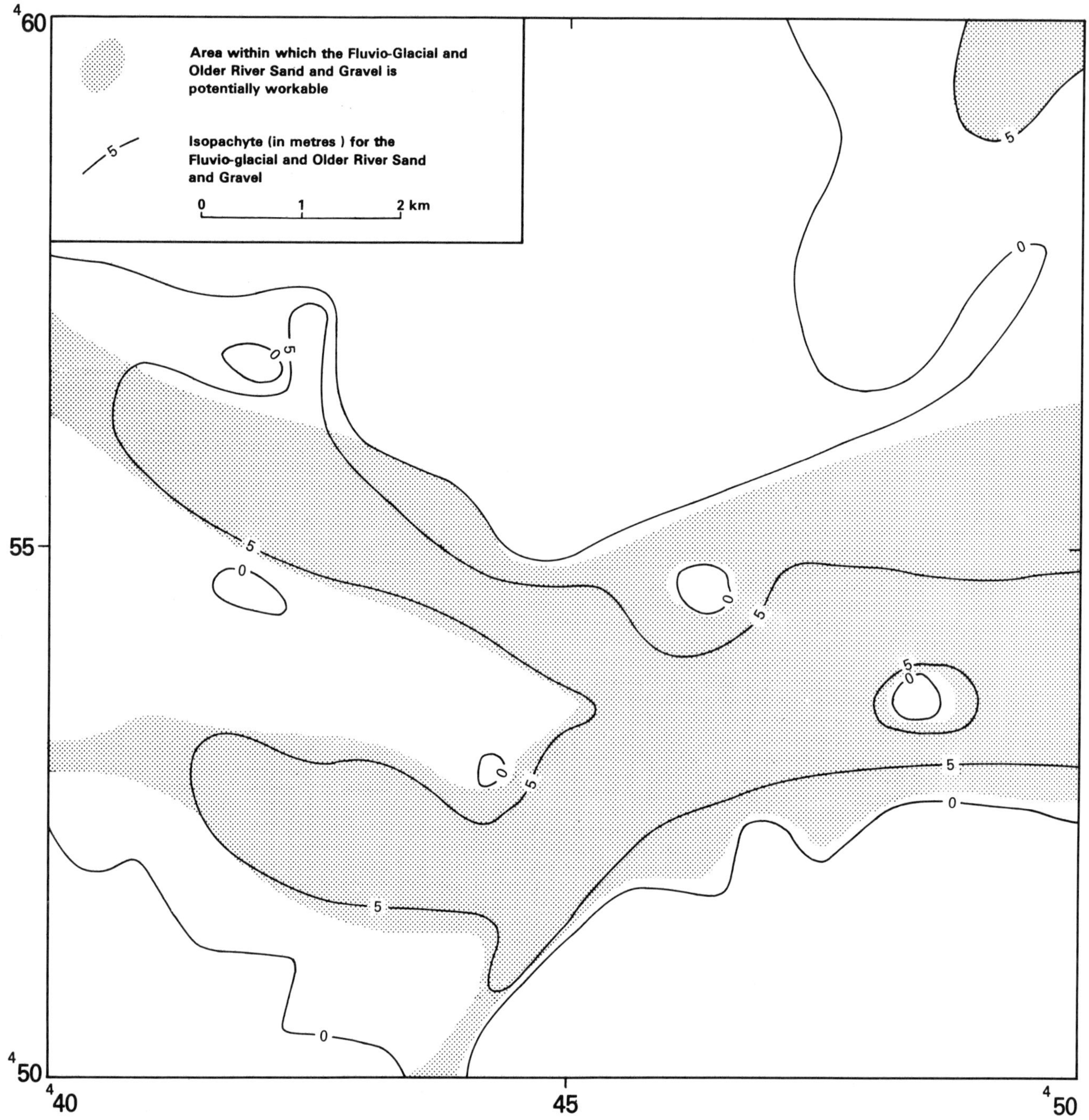

Figure 4 Isopachyte map of the Fluvio-glacial and Older River Sand and Gravel.

cent, sand 66 per cent and gravel 25 per cent does not accurately describe the distribution of individual gradings (see Figure 5). The triangular diagram shows two populations: a group of gravels and sandy gravels with a mean grading of fines 4 per cent, sand 54 per cent and gravel 42 per cent and a group of clean to 'very clayey' sands and pebbly sands with a mean grading of fines 15 per cent, sand 81 per cent and gravel 4 per cent. This division is reflected in the borehole records which frequently show 'clayey' sand lying upon sandy gravel, as in SE 29.

The gravel fraction usually shows approximately equal proportions of fine and coarse gravel with the coarse gravel fraction often slightly in excess. A small percentage of cobbles may also be present. Sandstone is the dominant component of the gravel fraction (Table 2), but it is normally associated with limestone. In some cases, however, limestone may be in excess of sandstone. Subordinate amounts of quartz, chert, ironstone and shale are also present in varying quantities.

Medium and fine sand are present in approximately equal percentages, though medium sand is frequently slightly in excess of the fine-sand grade. A small amount of coarse sand is normally present also.

Sandy Till This deposit is commonly composed of 'very clayey' pebbly sand with a mean grading of fines 28 per cent, sand 63 per cent and gravel 9 per cent. However, it also contains some 'clayey' or 'very clayey' sand and sandy gravel. The gravel fraction is composed of similar amounts of fine and coarse pebbles; cobbles are also occasionally present. The gravels are composed mainly of sandstone, generally with, subordinate amounts of limestone (Table 2); in addition small quantities of vein quartz, chert, basic igneous rocks, ironstone and shale may occur. The sand fraction is dominated by fine sand with lesser amounts of medium and coarse sand. Exceptionally, the percentage of medium sand may exceed that of fine sand.

Table 2 Lithological analyses of selected samples from the major mineral-bearing deposits.

Deposit	Borehole	Depth (m)	Percentage by weight						
			Sandstone	Limestone	Shale	Quartz	Chert	Basic igneous	Ironstone
Glacial Sand and Gravel	NW 25	0.0-4.0	64	32	0	1	2	0	1
	NW 25	7.3-10.2	47	49	trace	1	2	0	1
	NW 36	2.7-5.3	83	13	1	trace	1	1	1
	NE 11	1.5-7.8	70	25	0	trace	5	trace	trace
	SW 89	3.0-7.1	64	24	1	1	6	0	4
Sandy Till	NW 16	0.2-3.0	93	1	0	trace	trace	6	0
	NW 18	2.8-12.8	92	7	0	1	0	trace	0
	NW 24	0.2-3.2	98	0	0	1	0	1	0
	NW 35	0.6-2.8	85	2	trace	1	10	0	2
Fluvio-glacial and Older River Sand and Gravel	NW 32	15.7-23.2	72	23	0	2	2	0	1
	NW 33	19.3-24.3	88	3	1	5	2	0	1
	NE 15	12.5-21.5	79	14	0	1	3	0	3
	NE 16	14.7-21.2	79	11	0	1	4	0	5
	NE 31	12.0-15.1	94	2	trace	3	1	0	trace
	SW 71	8.8-11.9	79	2	trace	14	1	0	4
	SW 78	14.1-19.2	56	37	1	3	2	0	1
	SW 88	11.5-15.7	75	17	1	4	1	0	2
	SW 89	16.5-24.1	56	39	1	3	1	0	trace
	SE 23	7.8-16.0	90	6	trace	2	2	0	trace
	SE 25	11.0-13.6	91	1	0	6	1	0	1

Glacial Sand and Gravel The mean grading of this deposit is fines 12 per cent, sand 60 per cent and gravel 28 per cent. However, as with the Fluvio-glacial and Older River Sand and Gravel, the distribution of samples on the triangular diagram (Figure 5) appears to indicate the occurrence of two populations. The first of these is a group of clean to 'clayey' gravels and sandy gravels, with a mean grading of fines 6 per cent, sand 42 per cent and gravel 52 per cent. The other population consists of clean to 'clayey' sands and pebbly sands with a few 'very clayey' sandy gravels; the mean grading of this cluster is fines 18 per cent, sand 76 per cent and gravel 6 per cent.

The gravel contains approximately equal proportions of coarse and fine material, the coarse gravel fraction is, however, generally slightly in excess of the fine gravel; normally a small percentage of cobbles may be present. The major component of the gravel is sandstone, though rarely limestone may be present in equal amounts and exceptionally, limestone may exceed sandstone (Table 2). Small percentages of shale, quartz, chert, basic igneous rocks and ironstone may be present also.

The sand fraction is composed chiefly of fine sand with lesser amounts of medium and coarse sand.

Fluvio-glacial Terrace Deposits Only one borehole, SW 84, was drilled in this deposit; it proved 4.3 m of 'clayey' pebbly sand. The gravel fraction consists of fine and coarse pebbles without cobbles and is composed of sandstone with some limestone; small amounts of chert and quartz are present also. Medium-grained sand is the dominant member of the sand fraction with lesser amounts of fine and coarse sand.

Glacial Lake Deposits (Sand and Gravel) The lateral variation in the grade of this deposit is quite marked. Its mean grading of fines 17 per cent sand 75 per cent and gravel 8 per cent gives a misleading impression of the nature of its aggregate content, which varies from 'clayey' to 'very clayey' and from sand to gravel. Where gravel is present, fine gravel normally exceeds coarse gravel; exceptionally, as in borehole SW 77, the cobble content may reach 12 per cent. The gravels are composed of sandstone with limestone. The sand fraction is mainly fine sand with medium and some coarse grades.

River Terrace Deposits, undifferentiated The available data for this deposit are limited, since it was proved in only two IMAU boreholes. They yield a mean grading of fines 21 per cent, sand 57 per cent and gravel 22 per cent; approximately equal amounts of fine and coarse pebble gravel are present, and in borehole SE 27 cobbles amount to 2 per cent. The percentage of sandstone exceeds that of limestone in both of the boreholes. Fine sand exceeds medium sand in SE 27 but the converse is true for SW 74; both boreholes contain a small percentage of coarse sand.

Alluvium Clean to 'very clayey' pebbly sand dominates this deposit, except in borehole SW 72, which proved gravel. The mean grading for the deposit is fines 13 per cent, sand 71 per cent and gravel 16 per cent. Coarse gravel normally exceeds the percentage of fine gravel; cobbles are recorded only in borehole NE 28, where they amount to 3 per cent. The gravels are composed of approximately equal proportions of sandstone and limestone; a small percentage of chert is present also. Approximately equal amounts of medium and fine grains are found in the sand fraction with a lesser proportion of coarse sand.

THE MAP

The sand and gravel resource map is folded into the pocket at the end of this report. The base map is the Ordnance Survey 1:25 000 Outline Edition, on which the topography is shown by contours in grey, the geological data are in black and the mineral resource information in shades of red.

Geological data The geological boundary lines, symbols, etc., shown are taken from the geological map of this area, which was surveyed recently at the scale of 1:10 000. This information was obtained by detailed application of field mapping techniques by the field staff in the Institute's Yorkshire and East Midlands Unit. Borehole data, which include the stratigraphic relations, thicknesses and mean particle size distribution of the sand and gravel samples collected during the assessment survey, are also shown.

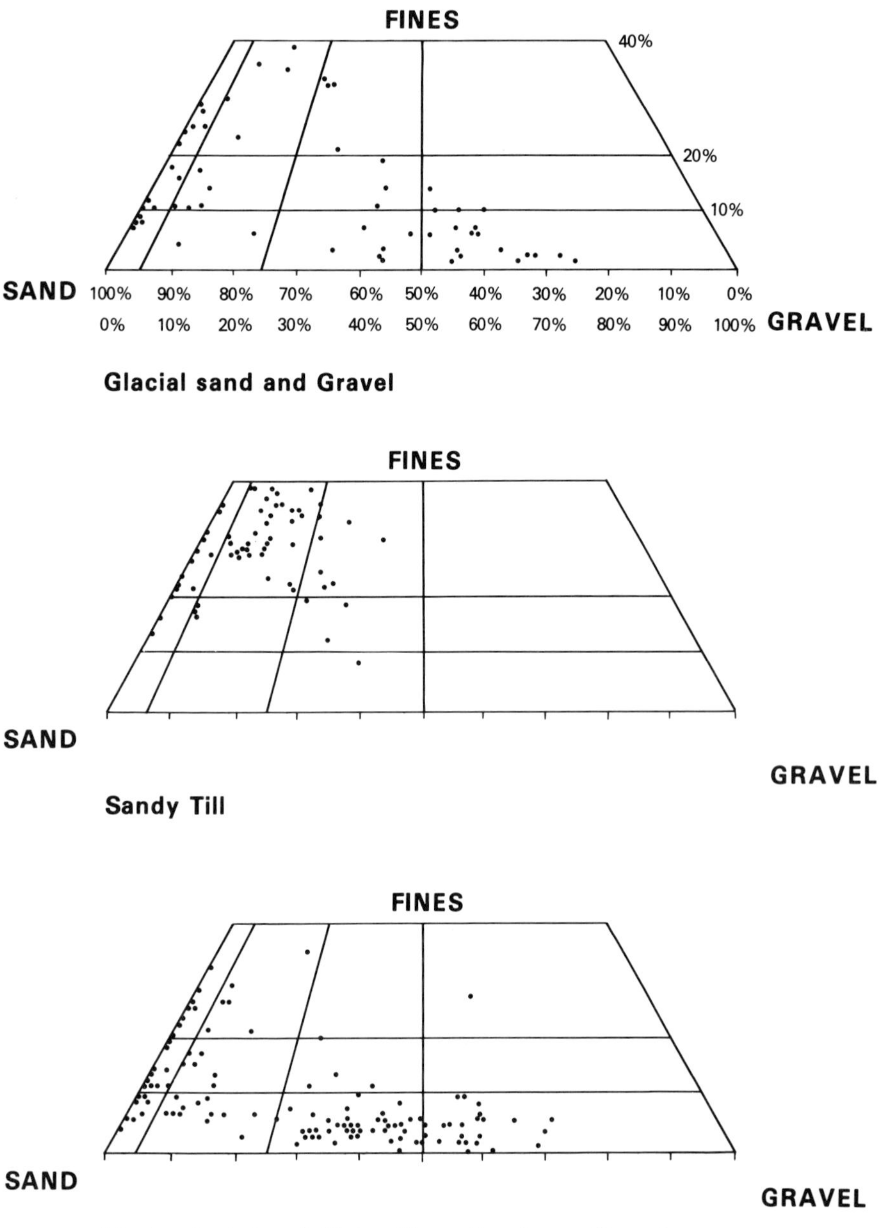

Figure 5 Diagrams comparing the particle size distribution of individual samples from the major mineral-bearing deposits.

The divisions in each diagram indicate the descriptive categories used in the classification of sand and gravel (see Appendix C).

The geological boundaries are the best interpretation of the information available at the time of survey. However it is inevitable, particularly with variable superficial deposits, that local irregularities and discrepancies will be revealed as new evidence from boreholes and excavations becomes available.

Mineral resource information: The mineral-bearing ground is divided into resource blocks (see Appendix A). Within a resource block the mineral is subdivided into areas where it is 'exposed' and areas where it is present beneath overburden. The mineral is identified as 'exposed' where the overburden, commonly consisting only of soil and subsoil, averages less than 1.0 metre in thickness. Beneath overburden the mineral may be continuous (or almost continuous) or discontinuous. The recognition of these categories is dependent upon the importance attached to the proportion of boreholes that did not find potentially workable sand and gravel and upon the distribution of barren boreholes within a block. The mineral is described as 'almost continuous' if it is present in 75 per cent or more of the boreholes in a resource block. The 'discontinuous' category has also been recognised on the present sheet.

Areas where bedrock crops out, where boreholes indicate absence of sand and gravel beneath cover and where sand and gravel beneath cover is interpreted to be not potentially workable, are uncoloured on the map; where appropriate, the relevant criterion is noted. In such cases it has been assumed that mineral is absent except in infrequent and relatively minor patches that can neither be outlined nor assessed quantitatively in the context of this survey. Areas of unassessed sand and gravel are indicated by a red stipple.

The area of the mineral-bearing ground is measured, where possible, from the mapped geological boundary lines. The whole of this area is considered as mineral-bearing, even though it may include small areas where sand and gravel is not present or is not potentially workable. Inferred boundaries have been inserted to delimit areas where sand and gravel beneath cover is interpreted to be not potentially workable or absent, or

Table 3 The sand and gravel resources of sheet SE 45: summary of statistical assessments.

Block and no. of sample points	Mineral-bearing deposits assessed	Area: Block km^2	Area: Mineral km^2	Mean thickness: Over-burden/depth of burial m	Mean thickness: Mineral m	Volume of sand and gravel m^3 × 10^6	Limits at the 95% probability level ±%	Limits at the 95% probability level ± m^3 × 10^6	Mean grading percentage: Fines $-\frac{1}{16}$ mm	Sand $+\frac{1}{16}$-4 mm	Gravel +4 mm
A (24)	All*	28.2	27.9	2.0	3.0	84	56	47	19	57	24
B (12)	FG & ORSG	21.1	12.4	9.4	7.0	87	25	22	12	64	24
(13)	All*	21.1	12.7	7.6	7.3	93	21	20	13	65	22
C (7)	FG & ORSG	28.5	10.1	12.3	6.2	63	35	22	7	68	25
(12)	All*	28.5	12.5	6.1	6.2	78	53	41	11	67	22
D (12)	FG & ORSG	10.8	10.6	12.3	5.7	60	25	15	9	66	25
(12)	All*	10.8	10.6	6.7	8.6	91	32	29	16	64	20
A-D(29)	FG & ORSG	88.6	33.1	11.2	6.3	209	15	31	9	66	25
(51)	All*	88.6	63.7	4.6	5.4	344	18	62	16	62	22

* All mineral excluding bedrock.

FG & ORSG Fluvio-glacial and Older River Sand and Gravel.

to indicate the approximate position of transitional boundaries between categories of deposit. Such boundaries (for which a distinctive zigzag symbol is used) are drawn primarily for the purpose of volume estimation. The symbol is intended to indicate an approximate location within a likely zone of occurrence rather than to represent the breadth of the zone, its size being determined only by cartographic considerations. For the purpose of measuring areas the centre line of the symbol is used.

RESULTS

The statistical results are summarised in Table 3. Fuller grading particulars are shown in Figures 6 to 10 and Tables 4 to 8.

For each block a statistical assessment of all potentially workable drift deposits is offered. A separate statistical assessment of the Fluvio-glacial and Older River Sand and Gravel is also offered for Resource Blocks B, C and D.

For the four resource blocks on sheet SE 45 the confidence limits at the 95 per cent probability level vary between 21 and 56 per cent (that is, on average nineteen out of every twenty sets of limits constructed in this way contain the true value for the volume of mineral). However, the true values are more likely to be nearer the figure estimated than the limits. Moreover, it is probable that in each block approximately the same percentage limits would apply for the estimated volume of a very much smaller parcel of ground (say, 100 hectares) containing similar sand and gravel deposits if results from the same number of sample points (as provided by, say, ten boreholes) were used in the calculation. Thus, if closer limits are needed for the quotation of reserves of part of a block, it can be expected that data from more than ten sample points will be required, even if the area is quite small. This point can be illustrated by considering the whole of the potentially workable sand and gravel in Resource Blocks A to D on this sheet. The volume (344 million m^3) can be estimated to limits of ± 18 per cent at the 95 per cent confidence level by a calculation based on data from 51 sample points spread across the four resource blocks.

However, it must be emphasised that the quoted volume of sand and gravel has no simple relationship with the amount that could be extracted in practice, since no allowance has been made in the calculations for any restraints (such as existing buildings and roads) on the use of land for mineral working.

NOTES ON THE RESOURCE BLOCKS

The resource block boundaries have been determined in relation to the distribution of the Fluvio-glacial and Older River Sand and Gravel, the major mineral-bearing deposit of the district. This deposit, though having a large areal distribution beneath overlying deposits, is confined to channels cut into the bedrock surface (Figures 3 and 4). Blocks B, C and D are partially or wholly underlain by Fluvio-glacial and Older River Sand and Gravel, though in places, due to the thickness of overlying material, the deposit may be not potentially workable in the terms of the limiting criteria (p. 1). Block A lies outside the channel system and consequently contains no deposits of Fluvio-glacial and Older River Sand and Gravel.

Block A

The major drift deposits in this block, which lies to the north and west of the village of Kirk Hammerton, are tills and glacial sand and gravel which are overlain by silts and clays in the east of the block.

Thickness and grading data from the 13 mineral-bearing IMAU boreholes are summarised in Table 5 and Figure 7. A further 11 IMAU borheoles failed to prove mineral, but their distribution is such that the area of barren ground cannot be outlined. The mineral in this block has been assigned therefore to the category of 'discontinuous spreads of mineral beneath overburden'. The mean thickness of mineral in this block is 3.0 m and its mean grading is fines 19 per cent, sand 57 per cent and gravel 24 per cent. However, the mean gradings of the two mineral-bearing deposits, Glacial Sand and Gravel and Sandy Till, differ considerably. The former has a mean grading of fines 12 per cent, sand 53 per cent and gravel 35 per cent, whereas the latter has a mean grading of fines 24 per cent, sand 66 per cent and gravel 10 per cent.

The thickness of overburden ranges up to 7.3 m, in borehole NW 27, with a mean thickness of only 2.0 m. The overburden is composed of either sandy, pebbly clays or thin soil, except in borehole NE 18 where it is a mottled laminated clay.

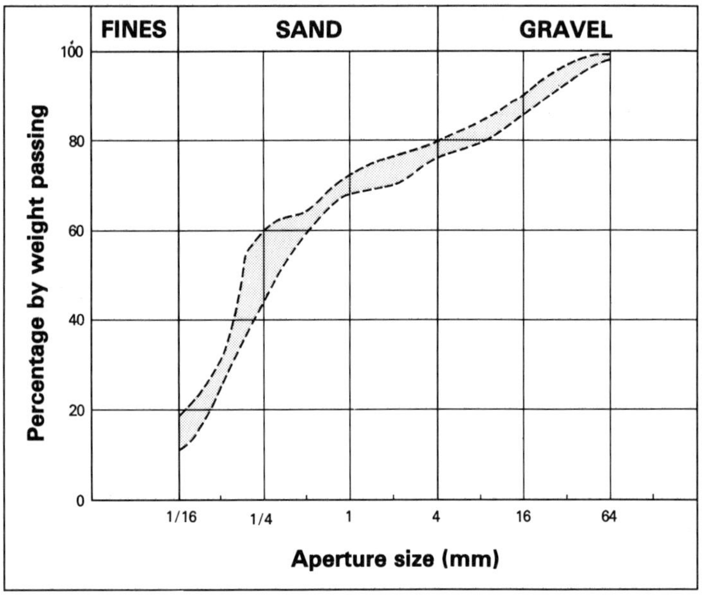

Figure 6 Envelope within which the mean grading curves for Resource Blocks A to D fall.

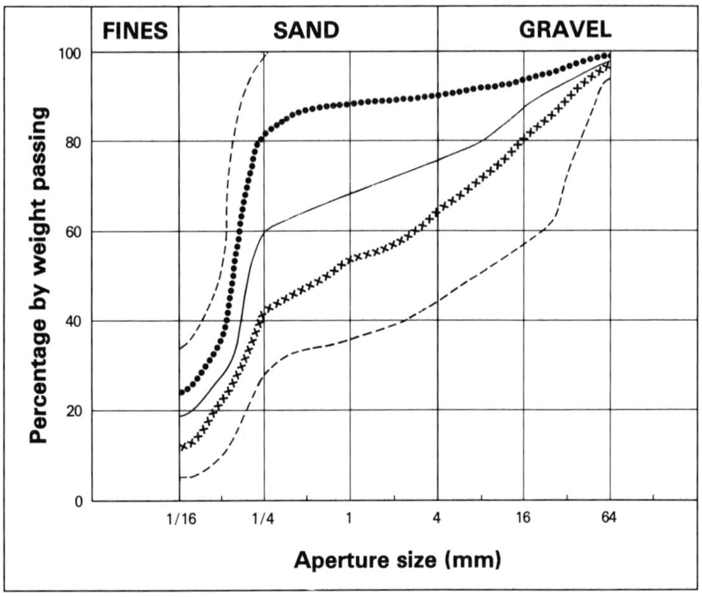

Figure 7 Particle size distribution of the mineral in Block A.

The continuous line represents the weighted mean grading of the block; the broken lines delimit the envelope within which the mean grading curves for individual boreholes fall; the lines depicted by crosses and circles are the weighted mean grading curves for the deposits of Glacial Sand and Gravel and Sandy Till, respectively.

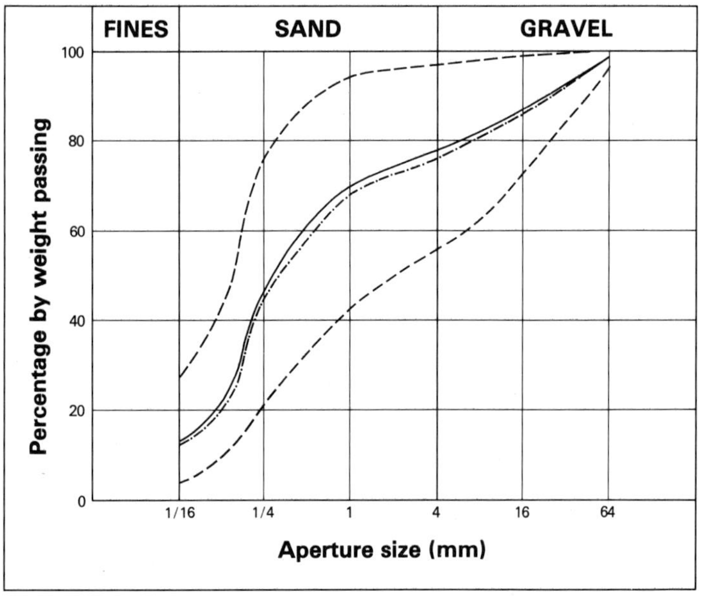

Figure 8 Particle size distribution of the mineral in Block B.

The continuous line represents the weighted mean grading of the block; the broken lines delimit the envelope within which the mean grading curves for individual boreholes fall; the broken line with dots represents the weighted mean grading of the Fluvio-glacial and Older River Sand and Gravel.

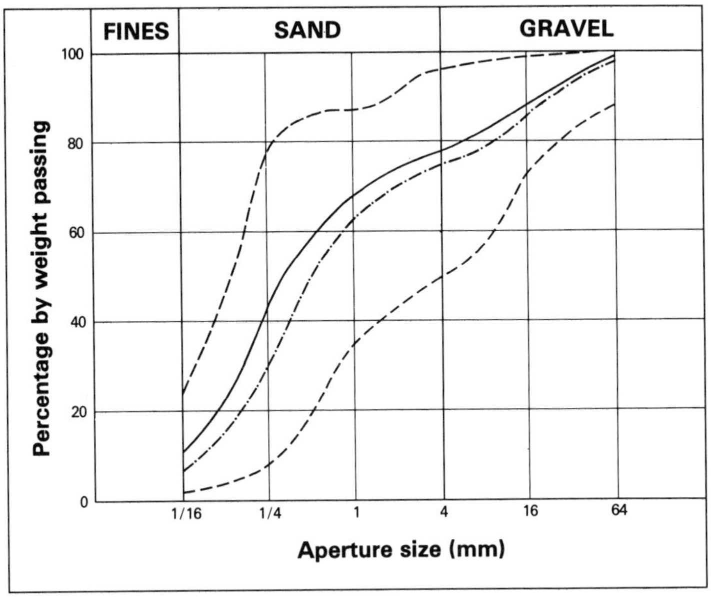

Figure 9 Particle size distribution of the mineral in Block C.

The continuous line represents the weighted mean grading of the block; the broken lines delimit the envelope within which the mean grading curves for individual boreholes fall. The broken line with dots represents the weighted mean grading of the Fluvio-glacial and Older River Sand and Gravel.

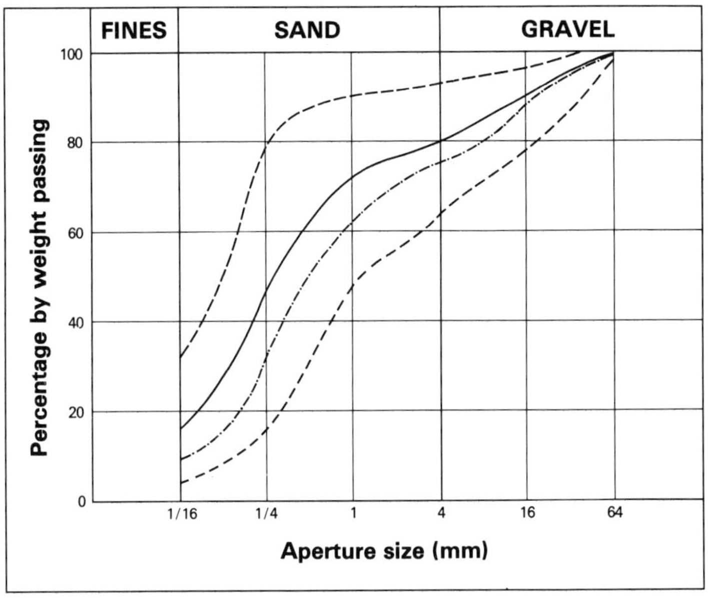

Figure 10 Particle size distribution of the mineral in Block D.

For explanation, see Figure 9.

The estimated volume of potentially workable sand and gravel present in the block is 84 million $m^3 \pm 56$ per cent. The wide confidence limits may be accounted for by the discontinuous and variable nature of the mineral bodies.

Block B

Fluvio-glacial and Older River Sand and Gravel, underlying other Drift deposits and occupying a buried channel incised in the bedrock surface, is the main mineral-bearing deposit in the block. It was proved in seventeen IMAU boreholes and one other borehole (SE 11), but in six of the IMAU boreholes the deposit is not potentially workable due to thick overburden. The other deposits that contain potentially workable sand and gravel in this block are the Glacial Lake Deposits and the Alluvium. Deposits assigned to Sandy Till proved in boreholes NE 33 and SE 24 are very similar in grading and composition to the upper, sandy part of the Fluvio-glacial and Older River Sand and Gravel found in adjacent boreholes. For the purposes of assessment therefore the Sandy Till has been included with the Fluvio-glacial and Older River Sand and Gravel.

Table 6 and Figure 8 summarise the thickness and grading data from the 12 IMAU boreholes that proved mineral in this block. The mean thickness of material overlying the potentially workable Fluvio-glacial and Older River Sand and Gravel is 9.4 m. This material, some of which is itself potentially workable, consists of tills, Glacial Lake Deposits, and Alluvium. The mean thickness of the potentially workable Fluvio-glacial and Older River Sand and Gravel is 7.0 m, and its mean grading is fines 12 per cent, sand 64 per cent and gravel 24 per cent. The estimated volume of potentially workable Fluvio-glacial and Older River Sand and Gravel is 87 million $m^3 \pm 25$ per cent.

The mean thickness of the combined potentially workable deposits in the block is 7.3 m and the mean grading is fines 13 per cent, sand 65 per cent and gravel 22 per cent. The mean thickness of overburden for the block as a whole is 7.6 m, and the estimated total volume of mineral in the block is 93 million $m^3 \pm 21$ per cent.

Table 4 Mean particle size distribution for the assessed thickness of sand and gravel in Resource Blocks A to D.

Resource block	Percentage by weight passing					
	$\frac{1}{16}$ mm	$\frac{1}{4}$ mm	1 mm	4 mm	16 mm	64 mm
A	19	60	69	76	86	98
B	13	46	70	78	87	99
C	11	44	68	78	88	98
D	16	47	72	80	90	99

Table 5 Block A: Data from IMAU boreholes.

Borehole	Recorded thickness		Mean grading percentage						
	Mineral m	Overburden/ Depth of burial m	Fines $-\frac{1}{16}$ mm	Fine sand $+\frac{1}{16}-\frac{1}{4}$ mm	Medium sand $+\frac{1}{4}$ -1 mm	Coarse sand +1 -4 mm	Fine gravel +4 -16 mm	Coarse gravel +16 -64 mm	Cobble gravel +64 mm
Sandy Till									
NW 16	2.8	0.2	20	46	7	4	7	16	0
NW 18	10.0	2.8	26	64	5	1	2	2	0
NW 35	2.2	0.6	20	41	11	4	6	18	0
NE 18	6.2*	2.3	22	62	10	2	3	1	0
NE 26	1.8	0.8	27	63	8	2	0	0	0
NE 30	4.0	2.0	29	43	9	2	7	5	5
Glacial Sand and Gravel									
NW 16	11.4+	13.4	6	10	11	17	25	29	2
NW 20	3.8	0.3	23	58	5	2	5	5	2
NW 25	10.2	0.0	5	30	18	11	15	15	6
NW 27	2.8	7.3	6	22	8	8	13	38	5
NW 31	7.0	2.1	34	39	8	4	5	5	5
NW 36	2.6	2.7	17	43	15	3	9	12	1
NE 11	6.3	1.5	3	27	16	18	21	14	1
NE 17	1.2	2.8	29	70	1	0	0	0	0

Potentially workable sand and gravel was absent in boreholes NW 17, NW 19, NW 21, NW 22, NW 23, NW 30, NW 34, NE 12, NE 19, NE 21, and NE 22.

* excluding a 3.4-m waste parting.

The plus sign indicates that the base of the deposit was not reached.

Table 6 Block B: Data from IMAU boreholes

Borehole	Recorded thickness		Mean grading percentage						
	Mineral m	Overburden/ Depth of burial m	Fines $-\frac{1}{16}$ mm	Fine sand $+\frac{1}{16}-\frac{1}{4}$ mm	Medium sand $+\frac{1}{4}$ -1 mm	Coarse sand +1 -4 mm	Fine gravel +4 -16 mm	Coarse gravel +16 -64 mm	Cobble gravel +64 mm
Alluvium									
NE 28	3.6	3.5	16	33	32	5	4	7	3
NW 32	4.8*	1.5	10	35	45	3	3	4	0
Glacial Lake Deposits									
NE 32	1.3	8.9	15	74	10	1	0	0	0
SE 23	1.0	1.7	37	33	13	7	7	3	0
Fluvio-glacial and Older River Sand and Gravel (including supposed **Sandy Till** in boreholes NE 33 and SE 24)									
NE 15	9.0	12.5	4	23	32	11	17	12	1
NE 16	6.5	14.7	8	31	24	12	14	10	1
NE 29	3.6	10.5	15	54	25	3	1	2	0
NE 32	2.2	10.6	5	19	14	6	16	37	3
NE 33	9.4	4.9	22	42	14	4	5	13	0
SE 23	8.2	7.8	15	35	12	7	9	20	2
SE 24	7.6	7.9	7	14	22	13	20	23	1
SE 25	4.8	8.8	9	36	14	5	9	24	3
SE 28	8.2	9.3	11	28	35	9	11	6	0
SE 29	10.9	9.6	12	36	30	7	6	8	1
SE 32	4.8	6.5	27	49	14	4	5	1	0

Potentially workable sand and gravel was absent in boreholes NE 13, NE 14, NE 20, NE 23, NE 24, NE 25, and NE 27.

* excluding a 1.5-m waste parting.

Table 7 Block C: Data from IMAU boreholes.

Borehole	Recorded thickness		Mean grading percentage						
	Mineral m	Overburden/ Depth of burial m	Fines $-\frac{1}{16}$ mm	Fine sand $+\frac{1}{16}-\frac{1}{4}$ mm	Medium sand $+\frac{1}{4}$ -1 mm	Coarse sand +1 -4 mm	Fine gravel +4 -16 mm	Coarse gravel +16 -64 mm	Cobble gravel +64 mm
Alluvium									
SW 72	1.9	0.9	9	11	20	14	19	27	0
SW 76	1.4	0.8	24	28	30	4	6	8	0
River Terrace Deposits									
SW 74	1.0	0.8	18	17	39	11	11	4	0
Glacial Lake Deposits									
SW 77	1.2	0.5	20	19	10	9	17	13	12
Fluvio-glacial Terrace Deposits									
SW 84	4.3	0.3	16	22	35	11	10	6	0
Glacial Sand and Gravel									
NW 28	2.5	6.7	5	14	11	12	22	27	9
SW 69	18.3*	0.5	15	62	10	3	5	4	1
Fluvio-glacial and Older River Sand and Gravel									
NW 28	8.3	12.1	4	19	32	11	15	19	0
NW 32	7.5	15.7	7	22	31	14	11	14	1
NW 33	9.4	14.9	6	38	33	8	7	8	0
SW 71	3.1	8.8	6	12	33	23	15	9	2
SW 73	4.7	11.8	10	24	53	9	3	1	2
SW 77	5.5	11.2	15	29	23	8	7	15	3
SW 81	4.9	11.5	2	6	27	15	24	25	1

Potentially workable sand and gravel was absent in boreholes NW 26, NW 29, SW 67, SW 75, SW 80, SW 85, SW 86, and SW 87.

* excluding a 0.3-m waste parting.

Table 8 Block D: Data from IMAU boreholes.

Borehole	Recorded thickness		Mean grading percentage						
	Mineral m	Overburden/ Depth of burial m	Fines $-\frac{1}{16}$ mm	Fine sand $+\frac{1}{16}-\frac{1}{4}$ mm	Medium sand $+\frac{1}{4}$ -1 mm	Coarse sand +1 -4 mm	Fine gravel +4 -16 mm	Coarse gravel +16 -64 mm	Cobble gravel +64 mm
River Terrace Deposits									
SE 27	3.4	1.6	23	34	14	5	10	12	2
Glacial Lake Deposits									
NE 31	5.0	0.2	13	66	17	3	1	0	0
Glacial Sand and Gravel									
SW 89	4.1	3.0	8	42	31	8	6	5	0
Sandy Till									
NE 31	1.0	7.5	27	47	11	4	5	6	0
SW 70	11.0	1.1	34	44	9	4	4	5	0
SE 31	4.0	1.7	30	48	11	3	6	2	0
SE 35	8.0	1.6	33	43	12	3	4	4	1
Fluvio-glacial and Older River Sand and Gravel									
NE 31	3.1	12.0	4	11	25	13	18	26	3
SW 70	8.0	13.9	13	46	13	5	9	13	1
SW 78	8.8	10.4	4	10	48	14	16	8	0
SW 83	6.0	15.8	7	11	30	17	17	18	0
SW 88	4.2	11.5	6	12	40	17	14	11	0
SW 89	7.6	16.5	6	27	35	10	13	9	0
SE 22	5.3	14.0	8	15	38	12	11	14	2
SE 26	5.2	13.1	4	12	32	16	14	22	0
SE 27	5.0	7.0	3	11	41	15	12	16	2
SE 31	7.3	8.5	30	44	13	7	2	4	0
SE 35	1.3	10.6	23	71	6	0	0	0	0

Potentially workable sand and gravel was absent in borehole SE 30.

Block C

Most of this block, which straddles the A1 trunk road between the southern border of the district near Deighton Whin [403 501] and just north of the A59 interchange [404 571], is underlain by Fluvio-glacial and Older River Sand and Gravel. This deposit is absent, however, to the south of the buried channel, in the vicinity of borehole SW 68, and also where the Sherwood Sandstone Group bedrock forms a small outcrop in the east bank [442 528] of the Nidd near Broad Oaks Farm.

The Fluvio-glacial and Older River Sand and Gravel extends over a larger area, 10.1 km^2, of mineral than any other deposit in the block and inferred boundaries have been used to separate the potentially workable portions of this deposit from the areas that are not mineral-bearing. The mean thickness of the potentially workable Fluvio-glacial and Older River Sand and Gravel is 6.2 m and its mean grading is fines 7 per cent, sand 68 per cent and gravel 25 per cent. The various glacial, lacustrine and fluvial deposits that overlie the channel-fill material have a mean thickness of 12.3 m, but this thickness includes some potentially workable material in places. The estimated volume of potentially workable Fluvio-glacial and Older River Sand and Gravel is 63 million $m^3 \pm 35$ per cent.

Boreholes NW 24, SW 68 and SW 79 prove sandy tills consisting of 'clayey' to 'very clayey' pebbly sand. However, it has proved impractical to separate the areas of potentially workable sandy till from the adjacent barren tills. Consequently the sandy till has been excluded from the assessment of this block.

The thickness of overburden above the assessed mineral in this block is 6.1 m. However, this figure is biased by the thickness of overburden above Fluvio-glacial and Older River Sand and Gravel and by the 6.7 m of till overlying mineral in borehole NW 28. If these are excluded the near-surface mineral in the block has a mean overburden thickness of 0.6 m.

Information from the 12 mineral bearing IMAU boreholes used in the assessment is summarised in Table 7 and Figure 9. The mean thickness of mineral in the block, excluding sandy till, is 6.2 m and the mean grading is fines 11 per cent sand 67 per cent and gravel 22 per cent. The estimated total volume of mineral is 78 million $m^3 \pm 53$ per cent.

Block D

This block is located to the south-west of Kirk Hammerton and to the north-west of Tockwith; it comprises an area of 10.8 km^2 which is underlain by a continuous spread of potentially workable Fluvio-glacial and Older River Sand and Gravel, except near a small inlier of Sherwood Sandstone Group at Tockwith Ness [463 546]. Other mineral-bearing deposits that occur in this block are Sandy Till, Glacial Sand and Gravel, Glacial Lake Deposits and River Terrace Deposits.

The potentially workable Fluvio-glacial and Older River Sand and Gravel was proved in 11 of the IMAU boreholes that lie within this block: the information from these boreholes is summarised in Table 8 and Figure 10. Additional information is provided by the record of a non-IMAU borehole, SE 9, which encountered 6.0 m of this deposit beneath 13.7 m of overburden. The beds that overlie the Fluvio-glacial and Older River Sand and Gravel have a mean thickness of 12.3 m and consist of glacial, lacustrine and fluvial deposits, some of which contain potentially workable material. The mean thickness of the Fluvio-glacial and Older River Sand and Gravel is 5.7 m and its mean grading is fines 9 per cent, sand 66 per cent and gravel 25 per cent. The estimated volume of potentially workable Fluvio-glacial and Older River Sand and Gravel present in the block is 60 million $m^3 \pm 25$ per cent.

The mean thickness of all potentially workable material is 8.6 m and its mean grading is fines 16 per cent, sand 64 per cent and gravel 20 per cent. The estimated total volume of mineral is 91 million $m^3 \pm 32$ per cent.

APPENDIX A

FIELD AND LABORATORY PROCEDURES

Trial and error during initial studies of the complex and variable glacial deposits of East Anglia and Essex showed that an absolute minimum of five sample points evenly distributed across the sand and gravel are needed to provide a worthwhile statistical assessment, but that, where possible, there should be not less than ten. Sample points are any points for which adequate information exists about the nature and thickness of the deposit and may include boreholes other than those drilled during the survey and exposures. In particular, the cooperation of sand and gravel operators ensures that boreholes are not drilled where reliable information is already available; although this may be used in the calculations, it is held confidentially by the Institute and cannot be disclosed.

The mineral shown on each 1:25 000 sheet is divided into resource blocks. The arbitrary size selected is a compromise to meet the aims of the survey by providing sufficient sample points in each block. As far as possible the block boundaries are determined by geological boundaries so that, for example, glacial and river terrace gravels are separated. Otherwise division is by arbitrary lines, which may bear no relationship to the geology. The blocks are drawn provisionally before drilling begins.

A reconnaissance of the ground is carried out to record any exposures and inquiries are made to ascertain what borehole information is available. Borehole sites are then selected to provide an even pattern of sample points at a density of approximately one per square kilometre. However, because broad trends are independently overlain by smaller-scale characteristically random variations, it is unnecessary to adhere to a square grid pattern. Thus such factors as ease of access and the need to minimise disturbance to land and the public are taken into account in siting the holes; at the same time it is necessary to guard against the possibility that ease of access (that is, the positions of roads and farms) may reflect particular geological conditions, which may bias the drilling results.

The drilling machine employed should be capable of providing a continuous sample representative of all unconsolidated deposits, so that the in-situ grading can be determined, if necessary, to a depth of 30 m (100 ft) at a diameter of about 200 mm (8 in), beneath different types of overburden. It should be reliable, quiet, mobile and relatively small (so that it can be moved to sites of difficult access). Shell and auger rigs have proved to be almost ideal.

The rigs are modified to enable deposits above the water table to be drilled 'dry', instead of with water added to facilitate the drilling, to minimise the amount of material drawn in from outside the limits of the hole. The samples thus obtained are representative of the in-situ grading, and satisfy one of the most important aims of the survey. Below the water table the rigs are used conventionally, although this may result in the loss of some of the fines fraction and the pumping action of the bailer tends to draw unwanted material into the hole from the sides or the bottom.

A continuous series of bulk samples is taken throughout the sand and gravel. Ideally samples are composed exclusively of the whole of the material encountered in the borehole between stated depths. However, care is taken to discard, as far as possible, material which has caved or has been pumped from the bottom of the hole. A new sample is commenced whenever there is an appreciable lithological change within the sand and gravel, or at every 1 m (3.3 ft) depth. The samples, each weighing between 25 and 45 kg (55 and 100 lb), are despatched in heavy-duty polythene bags to a laboratory for grading. The grading procedure is based on B.S. 1337 (British Standards Institution, 1967). Random checks of the accuracy of the grading are made in the Institute's laboratories.

All data, including mean grading analysis figures calculated for the total thickness of the mineral, are entered on standard record sheets, abbreviated copies of which are reproduced in Appendix E.

Detailed records may be consulted at the appropriate offices of the Institute, upon application to the Head, Industrial Minerals Assessment Unit.

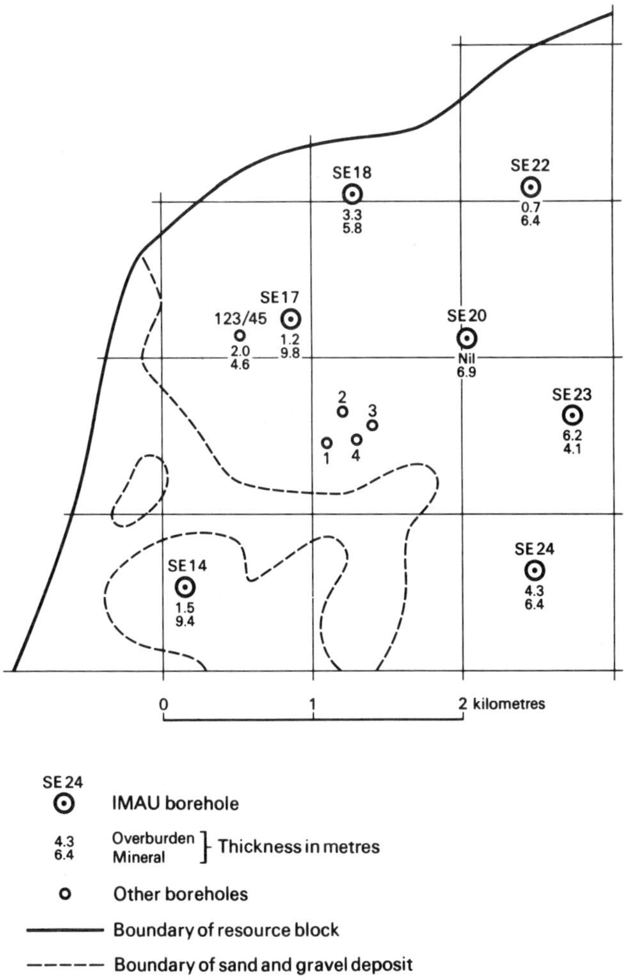

Example of resource block assessment: map of a fictitious block

APPENDIX B

STATISTICAL PROCEDURE

Statistical assessment

1 A statistical assessment is made of an area of mineral greater than 2 km^2, if there are at least five evenly spaced boreholes in the resource block (for smaller areas, see Paragraph 12 below).

2 The simple methods used in the calculations are consistent with the amount of data provided by the survey (Hull, 1981). Conventional symmetrical confidence limits are calculated for the 95 per cent probability level, that is, on average nineteen out of every twenty sets of limits constructed in this way contain the true value for the volume of mineral.

3 The volume estimate (V) for the mineral in a given block is the product of two variables, the sampled areas (A) and the mean thickness ($\bar{l}_m$) calculated from the individual thicknesses at the sample points. The standard deviations for these variables are related such that

$$S_V = \sqrt{(S_A^2 + S_{\bar{l}_m}^2)} \qquad [1]$$

4 The above relationship may be transposed such that

$$S_V = S_{\bar{l}_m}\sqrt{(1 + S_A^2 / S_{\bar{l}_m}^2)} \qquad [2]$$

From this it can be seen that as $S_A^2 / S_{\bar{l}_m}^2$ tends to 0, S_V tends to $S_{\bar{l}_m}$.

If, therefore, the standard deviation for area is small with respect to that for thickness, the standard deviation for volume approximates to that for mean thickness.

5 Given that the number of approximately evenly spaced sample points in the sampled area is n with mineral thickness measurements $l_{m1}, l_{m2}, \ldots l_{mn}$, then the best estimate of mean thickness, $\bar{l}_m$, is given by

$$\Sigma \; (l_{m1} + l_{m2} \ldots l_{mn}) / n .$$

For groups of closely spaced boreholes a discretionary weighting factor may be applied to avoid bias (see note on weighting below). The standard deviation for mean thickness $S_{\bar{l}_m}$, expressed as a proportion of the mean thickness, is given by

$$S_{\bar{l}_m} = (1/\bar{l}_m)\sqrt{[\Sigma(l_m - \bar{l}_m)^2/(n-1)]}$$

where l_m is any value in the series l_{m1} to l_{mn}.

6 The sampled area in each resource block is coloured pink on the map. Wherever possible, calculations relate to the mineral within mapped geological boundaries (which may not necessarily correspond to the limits of a deposit). Where the area is not defined by a mapped boundary, that is, where the boundary is inferred, a distinctive symbol is used. Experience suggests that the errors in determining area are small relative to those in thickness. The relationship $S_A / S_{\bar{l}_m} \leqq 0.\dot{3}$ is assumed in all cases. It follows from Equation [2] that

$$S_{\bar{l}_m} \leqq S_V \leqq 1.05\, S_{\bar{l}_m} \qquad [3]$$

7 The limits on the estimate of mean thickness of mineral, $L_{\bar{l}_m}$, may be expressed in absolute units

$\pm (t/\sqrt{n}) \times S_{\bar{l}_m}$ or as a percentage

$\pm (t/\sqrt{n}) \times S_{\bar{l}_m} \times (100/\bar{l}_m)$ per cent, where t is Student's t at the 95 per cent probability level for $(n-1)$ degrees of freedom, evaluated by reference to statistical tables. (In applying Student's t it is assumed that the measurements are distributed normally).

8 Values of t at the 95 per cent probability level for values of n up to 20 are as follows:

n	t	n	t
1	infinity	11	2.228
2	12.706	12	2.201
3	4.303	13	2.179
4	3.182	14	2.160
5	2.776	15	2.145
6	2.571	16	2.131
7	2.447	17	2.120
8	2.365	18	2.110
9	2.306	19	2.101
10	2.262	20	2.093

(from Table 12 in *Biometrika Tables for Statisticians*, Volume 1, Second Edition, Cambridge University Press, 1962). When n is greater than 20, 1.96 is used (the value of t when n is infinity).

9 In calculating confidence limits for volume, L_V, the following inequality, corresponding to Equation [3], is applied:

$$L_{\bar{l}_m} \leqq L_V \leqq 1.05\, L_{\bar{l}_m}.$$

10 In summary, for values of n between 5 and 20, L_V is calculated as

$$[(1.05 \times t)/\bar{l}_m] \times [\sqrt{\Sigma(l_m - \bar{l}_m)^2/n\ (n-1)}] \times 100$$

per cent,

and when n is greater than 20, as

$$[(1.05 \times 1.96)/\bar{l}_m] \times [\sqrt{\Sigma(l_m - \bar{l}_m)^2/n\ (n-1)}] \times 100$$

per cent.

11 The application of this procedure to a fictitious area is illustrated in the accompanying Figure and example of a block calculation.

Inferred assessment

12 If the sampled area of mineral in a resource block is between 0.25 km^2 and 2 km^2, an assessment is inferred on the basis of geological and topographical information, usually supported by the data from one or two boreholes. The volume of mineral is calculated as the product of the area, measured from field data, and the estimated thickness. Confidence limits are not calculated.

13 In some cases a resource block may include an area left uncoloured on the map, within which mineral (as defined) is interpreted to be generally absent. If there is reason to believe that some mineral may be present, an inferred assessment may be made.

14 No assessment is attempted for an isolated area of mineral less than 0.25 km^2.

15 Note on weighting The thickness of a deposit at any point may be governed solely by the position of the point in relation to a broad trend. However, most sand and gravel deposits also exhibit a random pattern of local, and sometimes considerable, variation in thickness. Thus the distribution of sample points needs to be only approximately regular and in estimating the mean thickness only simple weighting is necessary. In practice, equal weighting can often be applied to thicknesses at all sample points. If, however, there is a distinctly unequal distribution of points, bias is avoided by dividing the sampled area into broad zones, to each of which a value roughly proportional to its area is assigned. This value is then shared between the data points with the zone as the weighting factor.

Block calculation

Scale: 1:25 000
Block: Fictitious

Area
Block: 11.08 km^2
Mineral: 8.32 km^2

Mean thickness
Overburden: 2.5 m
Mineral: 6.5 m

Volume
Overburden: 21 million m^3
Mineral: 54 million m^3

Confidence limits of the estimate of mineral volume at the 95 per cent probability level: $\pm$ 20 per cent
That is, the volume of mineral (with 95 per cent probability): 54 $\pm$ 11 million m^3

Thickness estimate (measurements in metres)
l_o= overburden thickness l_m = mineral thickness

Sample point	Weighting w	Overburden l_o	Overburden wl_o	Mineral l_m	Mineral wl_m	Remarks
SE 14	1	1.5	1.5	9.4	9.4	IMAU boreholes
SE 18	1	3.3	3.3	5.8	5.8	
SE 20	1	nil	-	6.9	6.9	
SE 22	1	0.7	0.7	6.4	6.4	
SE 23	1	6.2	6.2	4.1	4.1	
SE 24	1	4.3	4.3	6.4	6.4	
SE 17	½	1.2	1.6	9.8	7.2	Hydrogeology Unit record
123/45	½	2.0		4.6		
1	¼	2.7	2.6	7.3	5.8	Close group of four boreholes (commercial)
2	¼	4.5		3.2		
3	¼	0.4		6.8		
4	¼	2.8		5.9		
Totals	$\Sigma w = 8$	$\Sigma wl_o = 20.2$		$\Sigma wl_m = 52.0$		
Means		$\overline{wl_o} = 2.5$		$\overline{wl_m} = 6.5$		

Calculation of confidence limits

wl_m	$\lvert(wl_m - \overline{wl_m})\rvert$	$(wl_m - \overline{wl_m})^2$
9.4	2.9	8.41
5.8	0.7	0.49
6.9	0.4	0.16
6.4	0.1	0.01
4.1	2.4	5.76
6.4	0.1	0.01
7.2	0.7	0.49
5.8	0.7	0.49

$\Sigma(wl_m - \overline{wl_m})^2 = 15.82$

$n = 8$

$t = 2.365$

L_V is calculated as

$$1.05\ (t/\overline{wl_m})\ \sqrt{[\Sigma(wl_m - \overline{wl_m})^2/\,n(n-1)]} \times 100$$

$$= 1.05 \times (2.365/6.5)\ \sqrt{[15.82/(8 \times 7)]} \times 100$$

$$= 20.3$$

$$\simeq 20 \text{ per cent.}$$

APPENDIX C

CLASSIFICATION AND DESCRIPTION OF SAND AND GRAVEL

For the purposes of assessing resources of sand and gravel a classification should take account of economically important characteristics of the deposit, in particular the absolute content of fines and the ratio of sand to gravel.

The terminology commonly used by geologists when describing sedimentary rocks (Wentworth, 1922) is not entirely satisfactory for this purpose. For example, Wentworth proposed that a deposit should be described as a 'gravelly sand' when it contains more sand than gravel and there is at least 10 per cent of gravel, provided that there is less than 10 per cent of material finer than sand ($< \frac{1}{16}$ mm) and coarser than pebbles ($>$ 64 mm in diameter). Because deposits containing more than 10 per cent fines are not embraced by this system, a modified binary classification based on Willman (1942) has been adopted.

When the fines content exceeds 40 per cent the material is considered to be not potentially workable and falls outside the definition of mineral. Deposits which contain 40 per cent fines or less are classified primarily on the ratio of sand to gravel but qualified in the light of the fines content, as follows: less than 10 per cent fines - no qualification; 10 per cent or more but less than 20 per cent fines - 'clayey'; 20 to 40 per cent fines - 'very clayey'.

The term 'clay' (as written, with single quote marks) is used to describe all material passing $\frac{1}{16}$ mm. Thus it has no mineralogical significance and includes particles falling within the size range of silt. The normal meaning applies to the term clay where it does not appear in single quotation marks.

The ratio of sand to gravel defines the boundaries between sand, pebbly sand, sandy gravel and gravel (at 19:1, 3:1 and 1:1).

Thus it is possible to classify the mineral into one of twelve descriptive categories (see the accompanying Figure). The procedure is as follows:

1 Classify according to the ratio of sand to gravel.
2 Describe the fines.

For example, a deposit grading 11 per cent gravel, 70 per cent sand and 19 per cent fines is classified as 'clayey' pebbly sand. This short description is included in the borehole log (see Appendix D)

Many differing proposals have been made for the classification of the grain size of sediments (Atterberg, 1905; Udden, 1914; Wentworth, 1922; Wentworth, 1935; Allen, 1936; Twenhofel, 1937; Lane and others, 1947). As Archer (1970a, b) has emphasised, there is a pressing need for a simple metric scale acceptable to both scientific and engineering interests, for which the class limit sizes correspond closely with certain marked changes in the natural properties of mineral particles. For example, there is an important change in the degree of cohesion between particles at about the $\frac{1}{16}$-mm size, which approximates to the generally accepted boundary between silt and sand. These and other requirements are met by a system based on Udden's geometric scale and a simplified form of Wentworth's terminology (see the accompanying table), which is used in the Report.

The fairly wide intervals in the scale are consistent with the general level of accuracy of the qualitative assessments of the resource blocks. Three sizes of sand are recognised, fine ($+\frac{1}{16}$ $-\frac{1}{4}$ mm), medium ($+\frac{1}{4}$ -1 mm) and coarse (+1 -4 mm). The boundary at 16 mm distinguishes a range of finer gravel (+4 -16 mm), often characterised by abundance of worn tough pebbles of vein quartz, from larger pebbles, often of notably different materials. The boundary at 64 mm distinguishes pebbles from cobbles. The term 'gravel' is used loosely to denote both pebble-sized and cobble-sized material.

The size distribution of borehole samples is determined by sieve analysis, which is presented by the laboratory as logarithmic cumulative curves (see, for example, British Standards Institution, 1967). In this report the grading is tabulated on the borehole record sheets (Appendix E), the intercepts corresponding with the simple geometric scale $\frac{1}{16}$ mm, $\frac{1}{4}$ mm, 1 mm, 4 mm, 16 mm and so on as required. Original sample grading curves are available for reference at the appropriate office of the Institute.

Each bulk sample is described, subjectively, by a geologist at the borehole site. Being based on visual examination, the description of the grading is inexact, the accuracy depending on the experience of the observer. The descriptions recorded are modified, as necessary, when the laboratory results become available.

The relative proportions of the rock types present in the gravel fraction are indicated by the use of the words 'and' or 'with'. For example, 'flint and quartz' indicates roughly equal proportions with neither constituent accounting for less than about 25 per cent of the whole; 'flint with quartz' indicates that flint is dominant and quartz, the principal accessory rock type, comprises 5 to 25 per cent of the whole. Where the accessory material accounts for less than 5 per cent of the whole, but is still readily apparent, the phrase 'with some' has been used. Rare constitutents are referred to as 'trace'.

The terms used in the field to describe the degree of rounding of particles, which is concerned with the sharpness of the edges and corners of a clastic fragment and not the shape (after Pettijohn, 1957), are as follows.

Angular: showing little or no evidence of wear; sharp edges and corners.

Subangular: showing definite effects of wear. Fragments still have their original form but edges and corners begin to be rounded off.

Subrounded: showing considerable wear. The edges and corners are rounded off to smooth curves. Orignal grain shape is still distinct.

Rounded: original faces almost completely destroyed, but some comparatively flat surfaces may still remain. All original edges and corners have been smoothed off to rather broad curves. Original shape is still apparent.

Well rounded: not original faces, edges or corners left. The entire surface consists of broad curves; flat areas are absent. The original shape is suggested by the present form of the grain.

Classification of gravel, sand and fines

Size limits	Grain-size description	Qualification	Primary classification
	Cobble		
64 mm			
		Coarse	Gravel
16 mm	Pebble		
		Fine	
4 mm			
		Coarse	
1 mm			
	Sand	Medium	Sand
$\frac{1}{4}$ mm			
		Fine	
$\frac{1}{16}$ mm			
	Fines (silt and clay)		Fines

I Gravel

II 'Clayey' gravel

III 'Very clayey' gravel

IV Sandy gravel

V 'Clayey' sandy gravel

VI 'Very clayey' sandy gravel

VII Pebbly sand

VIII 'Clayey' pebbly sand

IX 'Very clayey' pebbly sand

X Sand

XI 'Clayey' sand

XII 'Very clayey' sand

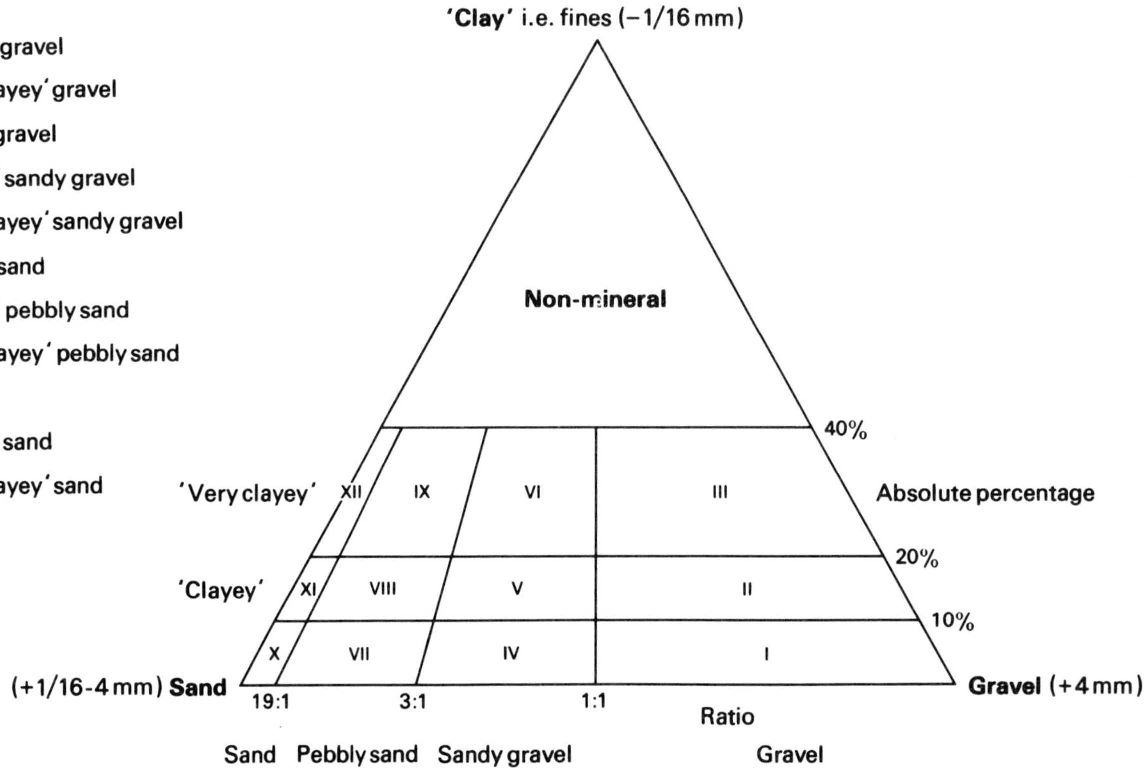

Diagram showing the descriptive categories used in the classification of sand and gravel

APPENDIX D

EXPLANATION OF THE BOREHOLE RECORDS

Annotated example

SE 45 NE 18 **4641 5853** **Divel Covert**[1] **Block A**

Surface level +17.5 m[2]
Water struck at +15.2 m[3]
October 1977[4]

Overburden 2.3 m
Mineral 1.5 m
Waste 3.4 m
Mineral 4.7 m
Bedrock 1.4 m+[5]

LOG

Geological classification	Lithology[6]	Thickness m	Depth m
	Soil	0.8	0.8
Glacial Lake Deposits	Clay, mottled grey, sandy in parts, laminated below 1.4 m	1.5	2.3
Sandy Till	**a** 'Very clayey' pebbly sand Gravel: fine, rounded sandstone and subangular to rounded limestone Sand: fine, quartz and lithic grains	1.5	3.8
Till	Clay, sandy, pebbly, brown	3.4	7.2
Sandy Till	**b** 'Very clayey' sand: fine, quartz with lithic grains, with some limestone gravel below 10.6 m	4.7	11.9
Sherwood Sandstone Group	Mudstone, reddish brown, weathered to 12.8 m	1.4+	13.3

GRADING[7]

	Mean for deposit percentages			Depth below surface (m)	percentages						
	Fines	Sand	Gravel		Fines	Sand			Gravel		
					$-\frac{1}{16}$	$+\frac{1}{16}-\frac{1}{4}$	$+\frac{1}{4}-1$	+1 -4	+4 -16	+16 -64	+64 mm
a	23	63	14	2.3-3.8	23	49	11	3	10	4	0
b	22	77	1	7.2-9.4	20	69	11	0	0	0	0
				9.4-10.6	26	62	10	2	0	0	0
				10.6-11.9	21	66	7	3	1	2	0
				Mean	22	66	10	1	0	1	0
a + b	22	74	4	Mean	22	62	10	2	3	1	0

The numbered paragraphs below correspond with the annotations given on the specimen record:

1 Location
The position of the borehole is generally referred to the nearest named locality on the 1:25 000 base map and the resource block in which it lies is stated.

2 Surface level
The surface level at the borehole site is given in metres above Ordnance Datum. All measurements were made in metres.

3 Groundwater conditions
If groundwater was present the level at which it was encountered is normally given (in metres above OD).

4 Type of drill and date of drilling
Unless other wise stated, all boreholes were drilled by a shell and auger rig using 152-m casing. The month and year of completion of the borehole are stated.

5 Thickness
All measurements were made in metres to the nearest 0.1 m. The plus sign (+) indicates that the base of the deposit was not reached during drilling.

6 Lithological description
When sand and gravel is recorded a general description based on the mean grading characteristics (for details see Appendix C) is followed by more detailed particulars. The description of other rocks is based on visual examination in the field. Where more than one mineral deposit is recognised, each is designated by a letter, e.g. **a**, **b**, etc.

7 Grading data
A continous series of bulk samples is taken throughout the thickness of sand and gravel. A new sample is commenced whenever there is an appreciable lithological change or at every 1 metre of depth.

For each bulk sample the percentages of fines ($-\frac{1}{16}$ mm), fine sand ($+\frac{1}{16}-\frac{1}{4}$), medium sand ($+\frac{1}{4}$ -1 mm), coarse sand (+1 -4 mm), fine gravel (+4 -16 mm) and coarse gravel (+16 mm) are stated. The mean grading of groups of samples making up an identified mineral horizon are also given in detail and, to the left, in summary. Where more than one horizon is recognised the mean grading for the whole of the mineral in the borehole is also given. Where necessary in calculating the mean grading, data for individual samples are weighted by the thickness represented.

Fully representative sampling of sand and gravel is difficult to achieve, particularly where groundwater levels are high. Comparison between boreholes and adjacent exposures suggests that in borehole samples the proportion of sand may be higher and the proportions of fines and coarse gravel (+16 mm) may be lower.

APPENDIX E

INDUSTRIAL MINERALS ASSESSMENT UNIT BOREHOLE RECORDS

SE 45 NW 16 **4097 5939** **Far Park** **Block A**

Surface level +65.6 m
Water not encountered
October 1977

Overburden 0.2 m
Mineral 2.8 m
Waste 10.4 m
Mineral 11.4 m+

LOG

Geological classification	Lithology	Thickness m	Depth m
	Soil	0.2	0.2
Sandy Till	**a** 'Very clayey' sandy gravel Gravel: coarse, well-rounded to angular, sandstone with some volcanics Sand: fine, quartz with some lithic grains Fines: thin seams of laminated clay	2.8	3.0
Laminated Clay closely associated with Glacial Deposits	Clay, Laminated with irregularly interbedded seams of silts and sands, brown	3.4	6.4
Till	Clay, sandy, pebbly, brown, thin seam of 'very clayey' pebbly sand 7.8 - 8.2 m	7.0	13.4
Glacial Sand and Gravel	**b** Gravel Gravel: coarse, angular to well-rounded limestone and angular to subangular sandstone with some volcanics Sand: coarse, angular quartz with angular to well rounded lithic grains	11.4+	24.8

GRADING

	Mean for deposit percentages			Depth below surface (m)	percentages						
	Fines	Sand	Gravel		Fines	Sand			Gravel		
					$-\frac{1}{16}$	$+\frac{1}{16}-\frac{1}{4}$	$+\frac{1}{4}-1$	+1 -4	+4 -16	+16 -64	+64 mm
a	20	57	23	0.2-2.0	21	46	9	5	8	11	0
				2.0-3.0	18	45	5	3	5	24	0
				Mean	20	46	7	4	7	16	0
b	6	38	56	13.4-14.3	10	7	13	19	32	19	0
				14.3-15.3	10	5	11	19	27	28	0
				15.3-16.9	14	16	9	17	13	31	0
				16.9-17.9	7	14	11	16	13	39	0
				17.9-19.1	6	19	13	17	18	27	0
				19.1-20.1	3	12	10	21	31	23	0
				20.1-21.4	2	9	10	13	30	26	10
				21.4-22.5	1	3	6	16	28	33	13
				22.5-23.5	1	3	8	14	31	43	0
				23.5-24.8	1	5	20	20	32	22	0
				Mean	6	10	11	17	25	29	2
a + b	8	42	50	Mean	8	17	10	15	21	27	2

SE 45 NW 17 **4154 5990** **Walls Close House** **Block A**

Surface level +58.7 m
Water not encountered
October 1977

Waste 18.2 m+

LOG

Geological classification	Lithology	Thickness m	Depth m
	Soil	0.3	0.3
Till	Clay, sandy, pebbly, sporadically interbedded seams of 'clayey' sand below 1.6 m	17.9+	18.2

SE 45 NW 18 **4272 5947** **Ox Close Farm** **Block A**

Surface level +55.3 m
Water not encountered
October 1977

Overburden 2.8 m
Mineral 10.0 m
Bedrock 0.3 m+

LOG

Geological classification	Lithology	Thickness m	Depth m
	Soil	0.6	0.6
Till	Clay, sandy, pebbly, brown	2.2	2.8
Sandy Till	'Very clayey' pebbly sand Gravel: fine and coarse, angular to subrounded, sandstone with some limestone Sand: fine, subangular to subrounded, quartz with some lithic grains Fines: thin seams of pebbly sandy clay between 4.1 m and 8.8 m, and below 10.1 m	10.0	12.8
Sherwood Sandstone Group	Sandstone, reddish brown	0.3+	13.1

GRADING

Mean for deposit percentages			Depth below surface (m)	percentages						
Fines	Sand	Gravel		Fines	Sand			Gravel		
				$-\frac{1}{16}$	$+\frac{1}{16}-\frac{1}{4}$	$+\frac{1}{4}-1$	+1 -4	+4 -16	+16 -64	+64 mm
26	70	4	2.8-4.1	13	81	5	1	0	0	0
			4.1-5.9	27	60	6	1	4	2	0
			5.9-6.9	36	42	4	2	5	11	0
			6.9-7.9	28	56	5	3	6	2	0
			7.9-8.8	30	60	4	2	4	0	0
			8.8-9.7	21	77	2	0	0	0	0
			9.7-10.8	27	61	3	2	1	6	0
			10.8-11.8	23	72	4	1	0	0	0
			11.8-12.8	30	63	7	0	0	0	0
			Mean	26	64	5	1	2	2	0

SE 45 NW 19 **4376 5903** **High Farm** **Block A**

Surface level +51.8 m
Water not encountered
October 1977

Waste 8.7 m
Bedrock 1.1 m+

LOG

Geological classification	Lithology	Thickness m	Depth m
	Soil	0.3	0.3
Till	Clay, sandy in part, mottled, with interbedded seams of pebbly sand, brown	1.9	2.2
Peat	Peat, clayey, pebbly	0.9	3.1
Glacial Sand and Gravel	'Clayey' sandy gravel Gravel: subrounded to rounded, tabular, limestone Sand: medium, quartz	0.6	3.7
Till	Clay, silty, pebbly, brown, becoming sandy towards base	3.0	6.7
Glacial Sand and Gravel	'Clayey' sandy gravel Gravel: coarse and fine, subrounded to rounded lime-Stone and sandstone with weathered sandstone bedrock Sand: fine, medium and coarse, angular, quartz with lithic grains	2.0	8.7
Sherwood Sandstone Group	Sandstone, reddish brown, weathered	1.1+	9.8

SE 45 NW 20 **4084 5836** **Allerton Park** **Block A**

Surface level +60.9 m
Water not encountered
August 1977

Overburden 0.3 m
Mineral 3.8 m
Waste 11.8 m
Bedrock 2.9 m+

LOG

Geological classification	Lithology	Thickness m	Depth m
	Soil	0.3	0.3
Glacial Sand and Gravel	'Very clayey' pebbly sand Gravel: fine and coarse, subrounded sandstone with angular to subangular limestone and some shale, chert and volcanics Sand: fine, quartz with some lithic grains Fines: sporadic thin seams of grey silt, laminated in part	3.8	4.1
Laminated Clay closely associated with Glacial Deposits	Silt, laminated, brown, with thin seams of silty clay and 'very clayey' sand	3.3	7.4
	Clay, laminated, with grey silt	3.5	10.9
Till	Clay, sandy, pebbly, reddish brown	5.0	15.9
Sherwood Sandstone Group	Sandstone, reddish brown, weathered	2.9+	18.8

GRADING

Mean for deposit percentages			Depth below surface (m)	percentages						
Fines	Sand	Gravel		Fines	Sand			Gravel		
				$-\frac{1}{16}$	$+\frac{1}{16}-\frac{1}{4}$	$+\frac{1}{4}-1$	+1 -4	+4 -16	+16 -64	+64 mm
23	65	12	0.3-1.1	10	28	8	7	16	24	7
			1.1-1.9	25	65	6	1	2	1	0
			1.9-3.1	30	60	4	2	3	1	0
			3.1-4.1	22	74	3	1	0	0	0
			Mean	23	58	5	2	5	5	2

SE 45 NW 21 **4145 5877** **Allerton Park** **Block A**

Surface level +54.1 m
Water not encountered
August 1977

Waste 14.6
Bedrock 1.1 m+

LOG

Geological classification	Lithology	Thickness m	Depth m
	Soil	0.3	0.3
Till	Clay, sandy, pebbly, dark grey	10.2	10.5
Glacial Sand and Gravel	Sandy Gravel Gravel: fine and coarse, sandstone and limestone Sand: medium and coarse, quartz with lithic grains	1.5	12.0
Till	Clay, pebbly, brown	2.6	14.6
Sherwood Sandstone Group	Sandstone, reddish brown, weathered to 15.3 m	1.1+	15.7

SE 45 NW 22 **4270 5834** **Willow Hole** **Block A**

Surface level +56.5 m
Water not encountered
October 1977

Waste 11.0 m
Bedrock 0.5 m+

LOG

Geological classification	Lithology	Thickness m	Depth m
	Soil	0.2	0.2
Till	Clays, interbedded with silt and sand, laminated in part with sporadic pebbles	4.7	4.9
	Clay, pebbly, brown, sandy below 6.6 m	5.0	9.9
Glacial Sand and Gravel	Pebby sand Gravel: fine and coarse, weathered sandstones Sand: coarse, angular, quartz and lithic grains	1.0	10.9
Till	Clay, pebbly, reddish brown	0.1	11.0
Sherwood Sandstone Group	Mudstone, reddish brown, weathered	0.5+	11.5

SE 45 NW 23 **4444 5833** **Whixley** **Block A**

Surface level +52.1 m
Water not encountered
October 1977

Waste 10.4 m
Bedrock 1.0 m+

LOG

Geological classification	Lithology	Thickness m	Depth m
	Made ground	0.7	0.7
Till	Clay, sandy, pebbly, brown, reddish brown below 8.1 m	9.7	10.4
Sherwood Sandstone Group	Sandstone and mudstone, reddish brwon, weathered to 11.1 m	1.0+	11.4

SE 45 NW 24 **4070 5724** **Ox Close** **Block A**

Surface level +42.3 m
Water struck at +40.1 m
October 1977

Overburden 0.2 m
Mineral 3.0 m
Waste 19.9 m
Bedrock 1.7 m+

LOG

Geological classification	Lithology	Thickness m	Depth m
	Soil	0.2	0.2
Sandy Till	'Clayey' pebbly sand Gravel: coarse, rounded, sandstone Sand: fine, angular quartz with rounded lithic grains	3.0	3.2
Clay associated with Fluvio-glacial and Older River Sand and Gravel	Silt and clay, laminated, brown, thin seams of 'clayey' and 'very clayey' sand below 6.0 m	16.1	19.3
Fluvio-glacial and Older River Sand and Gravel	'Clayey' sand: medium, quartz with lithic grains, and some sandstone and limestone gravel	3.8	23.1
Sherwood Sandstone Group	Sandstone, reddish brown, weathered to 24.7 m	1.7+	24.8

GRADING

Mean for deposit percentages			Depth below surface (m)	percentages						
Fines	Sand	Gravel		Fines	Sand			Gravel		
				$-\frac{1}{16}$	$+\frac{1}{16}-\frac{1}{4}$	$+\frac{1}{4}-1$	+1 -4	+4 -16	+16 -64	+64 mm
19	70	11	0.2-1.1	12	26	26	7	12	17	0
			1.1-2.2	27	57	8	1	2	5	0
			2.2-3.2	16	78	6	0	0	0	0
			Mean	19	55	13	2	4	7	0

SE 45 NW 25 **4167 5751** **Allerton Grange** **Block A**

Surface level +57.0 m
Water not encountered
December 1977

Mineral 10.2 m+

LOG

Geological classification	Lithology	Thickness m	Depth m
Glacial Sand and Gravel	**a** Gravel Gravel: fine and coarse, angular to subrounded, equant and tabular sandstone and subangular to rounded limestone with some chert Sand: medium and fine, subangular to subrounded, equant, quartz with angular to subrounded lithic grains	4.0	4.0
	b Sand: fine, subangular to subrounded, equant, quartz with lithic grains	3.3	7.3
	c Sandy gravel Gravel: coarse, subrouded to rounded, tabular to equant, limestone and subangular to subrounded, sandstone, with some angular chert Sand: fine, angular to subrounded, equant, quartz with lithic grains	2.9+	10.2

Borehole abandoned against boulder

GRADING

	Mean for deposit percentages			Depth below surface (m)	percentages						
	Fines	Sand	Gravel		Fines	Sand			Gravel		
					$-\frac{1}{16}$	$+\frac{1}{16}-\frac{1}{4}$	$+\frac{1}{4}-1$	+1 -4	+4 -16	+16 -64	+64 mm
a	3	47	50	0.0-2.0	3	18	10	8	16	22	23
				2.0-3.0	3	20	22	13	19	23	0
				3.0-4.0	3	7	28	28	28	6	0
				Mean	3	16	17	14	20	19	11
b	8	91	1	4.0-5.3	7	46	30	17	0	0	0
				5.3-6.3	8	64	23	4	1	0	0
				6.3-7.3	10	58	24	6	2	0	0
				Mean	8	55	26	10	1	0	0
c	4	38	58	7.3-8.7	6	23	15	8	22	26	0
				8.7-9.7	2	13	8	10	28	30	9
				9.7-10.2	2	13	8	10	28	30	9
				Mean	4	18	11	9	25	28	5
a+b+c	5	59	36	Mean	5	30	18	11	15	15	6

SE 45 NW 26 **4251 5723** **Hew Beck** **Block C**

Surface level +39.8 m
Water struck at +28.7 m
August 1977

Waste 25.0 m+

LOG

Geological classification	Lithology	Thickness m	Depth m
	Soil	0.5	0.5
Till	Clay, silty, pebbly, becoming sandy at 3.0 m, brown becoming reddish brown at 2.5 m	10.6	11.1
Glacial Sand and Gravel	'Clayey' gravel Gravel: fine, coarse and cobbles, rounded to well rounded sandstone, with some subangular chert Sand: fine, quartz with lithic grains	1.4	12.5
Till	Clay, sandy, pebbly, brown, thin seam of sandy silt at 14.2 m	1.8	14.3
Glacial Sand and Gravel	'Clayey' pebbly sand Gravel: fine and coarse, sandstone and quartzite with some shale Sand: fine, quartz and lithic grains	3.6	17.9
Clay associated with Fluvio-glacial and Older River Sand and Gravel	Clay, silty, laminated, brown	2.2	20.1
Fluvio-glacial and Older River Sand and Gravel	Gravel Gravel: coarse, subangular to rounded, sandstone and quartzite with limestone and angular chert Sand: fine, medium and coarse, angular to subangular quartz with lithic grains.	4.9+	25.0

SE 45 NW 27 **4368 5771** **Manor Farm** **Block A**

Surface level +42.1 m
Water struck at +34.8 m
October 1977

Overburden 7.3 m
Mineral 2.8 m
Bedrock 0.6 m+

LOG

Geological classification	Lithology	Thickness m	Depth m
	Soil	0.7	0.7
Till	Clay, sandy, pebbly, brown	6.6	7.3
Glacial Sand and Gravel	Gravel Gravel: coarse, subangular to rounded sandstone and limestone, with shale Sand: fine, quartz with lithic grains	2.8	10.1
Sherwood Sandston Group	Sandstone	0.6+	10.7

GRADING

Mean for deposit percentages			Depth below surface (m)	percentages						
Fines	Sand	Gravel		Fines	Sand			Gravel		
				$-\frac{1}{16}$	$+\frac{1}{16}-\frac{1}{4}$	$+\frac{1}{4}-1$	+1 -4	+4 -16	+16 -64	+64 mm
6	38	56	7.3-10.1	6	22	8	8	13	38	5

SE 45 NW 28 **4119 5612** **Green Dick Wood** **Block C**

Surface level +29.4 m
Water struck at +22.7 m
September 1977

Overburden 6.7 m
Mineral 2.5 m
Waste 2.9 mm
Mineral 8.3 m
Bedrock 0.8 m+

LOG

Geological classification	Lithology	Thickness m	Depth m
	Soil	0.3	0.3
Till	Clay, sandy, pebbly, brown	6.4	6.7
Glacial Sand and Gravel	**a** Gravel Gravel: coarse and fine, subangular to rounded limestone and subrounded to rounded sandstone with shale Sand: fine, medium and coarse, angular quartz with lithic grains	2.5	9.2
Till	Clay, sandy, pebbly, dark grey	1.8	11.0
Clay associated with Fluvio-glacial and Older River Sand and Gravel	Clay, laminated, grey, with silt parting	1.1	12.1
Fluvio-glacial and Older River Sand and Gravel	**b** Sandy gravel Gravel: coarse, rounded, sandstone with limestone and some shale and quartz Sand: medium, angular quartz with lithic grains	8.3	20.4
Sherwood Sandstone Group	Sandstone, reddish brown, weathered	0.8+	21.2

GRADING

	Mean for deposit percentages			Depth below surface (m)	percentages						
	Fines	Sand	Gravel		Fines	Sand			Gravel		
					$-\frac{1}{16}$	$+\frac{1}{16}-\frac{1}{4}$	$+\frac{1}{4}-1$	+1 -4	+4 -16	+16 -64	+64 mm
a	5	37	58	6.7-7.6	7	16	10	12	24	31	0
				7.6-8.7	2	8	7	12	27	26	18
				8.7-9.2	7	25	18	13	6	26	5
				Mean	5	14	11	12	22	27	9
b	4	62	34	12.1-13.2	6	15	45	13	15	6	0
				13.2-14.3	5	36	47	8	3	1	0
				14.3-15.3	5	26	46	8	9	6	0
				15.3-16.3	2	18	26	10	16	28	0
				16.3-17.2	0	3	23	13	35	26	0
				17.2-18.3	1	6	22	14	24	33	0
				18.3-18.9	2	6	15	8	13	51	5
				18.9-20.4	6	29	28	14	8	15	0
				Mean	4	19	32	11	15	19	0
a + b	4	57	39	Mean	4	18	27	12	16	21	2

SE 45 NW 29 **4192 5674** **Low Plantation** **Block C**

Surface level +31.8 m
Water struck at +21.3 m
August 1977

Waste 12.6 m
Bedrock 7.7 m+

LOG

Geological classification	Lithology	Thickness m	Depth m
	Soil	0.3	0.3
Till	Clay, sandy, pebbly, mottled to 3.0 m then brown	10.2	10.5
Glacial Sand and Gravel	Gravel Gravel: fine and coarse, subangular to well-rounded sandstone and angular to subangular limestone Sand: coarse, quartz and lithic grains	1.3	11.8
Till	Clay, pebbly, dark grey, with lenses of brown silt	0.8	12.6
Sherwood Sandstone Group	Sandstone, brown, weathered to 18.9 m	7.7+	20.3

SE 45 NW 30 **4297 5621** **Gelsthorpe** **Block A**

Surface level +31.6 m
Water level not recorded
August 1977

Waste16.0 m
Bedrock 4.6 m+

LOG

Geological classification	Lithology	Thickness m	Depth m
	Soil	0.3	0.3
Till	Clay, sandy, silt, pebbly, mottled, dark grey below 2.9 m, brown below 13.3 m, thin seam of 'very clayey' pebbly sand at 15.0 m	15.7	16.0
Sherwood Sandstone Group	Sandstone, weathered to 19.7 m	4.6+	20.6

SE 45 NW 31 **4336 5656** **Whixley Lodge** **Block A**

Surface level +36.1 m
Water struck at +22.2 m
December 1977

Overburden 2.1 m
Mineral 7.0 m
Bedrock 5.9 m+

LOG

Geological classification	Lithology	Thickness m	Depth m
	Soil	0.5	0.5
Till	Clay, sandy, pebbly, brown	1.6	2.1
Glacial Sand and Gravel	'Very clayey' pebbly sand Gravel: fine, coarse and cobble, subrounded to rounded, equant, sandstone with limestone Sand: fine, subangular to subrounded, equant, quartz with lithic grains	7.0	9.1
Sherwood Sandstone Group	Sandstone, brown	5.9+	15.0

GRADING

Mean for deposit percentages			Depth below surface (m)	percentages						
Fines	Sand	Gravel		Fines	Sand			Gravel		
				$-\frac{1}{16}$	$+\frac{1}{16}-\frac{1}{4}$	$+\frac{1}{4}-1$	+1 -4	+4 -16	+16 -64	+64 mm
34	51	15	2.1-3.1	39	39	7	5	5	5	0
			3.1-4.1	32	37	7	4	4	3	13
			4.1-5.1	33	38	8	3	6	6	6
			5.1-6.1	35	41	9	4	3	3	5
			6.1-7.1	36	45	9	4	4	2	0
			7.1-8.1	32	37	8	3	7	6	7
			8.1-9.1	32	38	8	3	7	12	0
			Mean	34	39	8	4	5	5	5

SE 45 NW 32 **4182 5534** **Whixley Lodge** **Block C**

Surface level +29.5 m
Water struck at +20.1 m
December 1977

Overburden 15.7
Mineral 7.5 m
Bedrock 0.8 m+

LOG

Geological classification	Lithology	Thickness m	Depth m
	Soil	0.3	0.3
Till	Clay, sandy, pebbly, brown, sporadic thin seam of fine-grained sand	15.4	15.7
Fluvio-glacial and Older River Sand and Gravel	Sandy Gravel Gravel: coarse, subangular to subrounded, equant, sandstone with subrounded to well-rounded, equant limestone and some angular chert and rounded quartz Sand: medium, angular to rounded, equant quartz with lithic grains	7.5	23.2
Sherwood Sandstone Group	Sandstone, brown	0.8+	24.0

GRADING

Mean for deposit percentages			Depth below surface (m)	percentages						
Fines	Sand	Gravel		Fines	Sand			Gravel		
				$-\frac{1}{16}$	$+\frac{1}{16}-\frac{1}{4}$	$+\frac{1}{4}-1$	+1 -4	+4 -16	+16 -64	+64 mm
7	67	26	15.7-16.8	14	43	35	8	0	0	0
			16.8-18.9	5	17	38	16	12	12	0
			18.9-19.9	4	17	24	7	14	32	2
			19.9-21.0	4	26	20	13	16	21	0
			21.0-22.0	2	13	46	19	11	9	0
			22.0-23.2	10	21	19	16	13	21	0
			Mean	7	22	31	14	11	14	1

SE 45 NW 33 **4258 5541** **Syke Dike** **Block C**

Surface level +29.0 m
Water struck at +25.2 m
January 1978

Overburden 14.9 m
Mineral 9.4 m
Bedrock 0.7 m+

LOG

Geological classification	Lithology	Thickness m	Depth m
	Soil	0.4	0.4
Till	Clay, sandy, pebbly, brown	14.1	14.5
Clay associated with Fluvio-glacial and Older River Sand and Gravel	Clay, laminated with partings of silt and fine-grained sand	0.4	14.9
Fluvio-glacial and Older River Sand and Gravel]	**a** Sand: fine, subangular to subrounded, equant, quartz with lithic grains	4.4	19.3
	b Sandy Gravel Gravel: coarse and fine, subangular to rounded, equant, sandstone with some rounded quartz and limestone, and angular chert Sand: medium, angular to rounded equant quartz with lithic grains	5.0	24.3
Sherwood Sandstone Group	Sandstone, reddish brown	0.7+	25.0

GRADING

	Mean for deposit percentages			Depth below surface (m)	percentages						
	Fines	Sand	Gravel		Fines	Sand			Gravel		
					$-\frac{1}{16}$	$+\frac{1}{16}-\frac{1}{4}$	$+\frac{1}{4}-1$	+1 -4	+4 -16	+16 -64	+64 mm
a	9	90	1	14.9-17.3	11	66	22	1	0	0	0
				17.3-18.3	9	52	36	2	1	0	0
				18.3-19.3	5	27	64	3	1	0	0
				Mean	9	54	34	2	1	0	0
b	4	69	27	19.3-20.4	6	14	29	11	20	20	0
				20.4-21.5	3	14	43	9	11	20	0
				21.5-22.5	2	18	33	15	15	17	0
				22.5-23.1	2	26	22	20	19	11	0
				23.1-24.3	6	40	30	17	5	2	0
				Mean	4	23	32	14	13	14	0
a + b	6	79	15	Mean	6	38	33	8	7	8	0

SE 45 NW 34 **4376 5581** **Cattal Grange** **Block A**

Surface level + 31.1 m
Water not encountered
December 1977

Waste 18.0 m+

LOG

Geological classification	Lithology	Thickness m	Depth m
	Soil	0.4	0.4
Till	Clay, sandy, pebbly, brown, cobbles found at 12.0 m	17.6+	18.0

SE 45 NW 35 **4433 5598** **Maltkiln** **Block A**

Surface level +34.4 m
Water not encountered
December 1977

Overburden 0.6 m
Mineral 2.2 m
Bedrock 0.1 m+

LOG

Geological classification	Lithology	Thickness m	Depth m
	Soil	0.6	0.6
Sandy Till	'Very clayey' sandy gravel Gravel: coarse, subangular to rounded, equant, sandstone with some, angular chert, limestone and ironstone Sand: fine, subangular to rounded, equant quartz with lithic grains	2.2	2.8
Sherwood Sandstone Group	Sandstone, brown	0.1+	2.9

GRADING

Mean for deposit percentages			Depth below surface (m)	percentages						
Fines	Sand	Gravel		Fines	Sand			Gravel		
				$-\frac{1}{16}$	$+\frac{1}{16}-\frac{1}{4}$	$+\frac{1}{4}-1$	+1 -4	+4 -16	+16 -64	+64 mm
20	56	24	0.6-1.6	22	38	11	4	4	21	0
			1.6-2.8	19	43	12	4	6	16	0
			Mean	20	41	11	4	6	18	0

SE 45 NW 36 **4442 5528** **Lingfield Cottage** **Block A**

Surface level +28.7 m
Water struck at +24.3 m
December 1977

Overburden 2.7 m
Mineral 2.6 m
Bedrock 1.2 m+

LOG

Geological classification	Lithology	Thickness m	Depth m
	Soil	0.4	0.4
Glacial Sand and Gravel	'Very clayey' pebbly sand Gravel: coarse and fine, rounded to well-rounded sandstone Sand: fine, subrounded, quartz with lithic grains	0.9	1.3
Till	Clay, sandy, pebbly, brown	1.4	2.7
Glacial Sand and Gravel	'Clayey' sandy gravel Gravel: coarse, angular to subrounded, equant, sandstone with subangular to subrounded limestone Sand: fine, subangular to rounded, equant, quartz with some lithic grains.	2.6	5.3
Sherwood Sandstone Group	Sandstone, brown	1.2+	6.5

GRADING

Mean for deposit percentages			Depth below surface (m)	percentages						
Fines	Sand	Gravel		Fines	Sand			Gravel		
				$-\frac{1}{16}$	$+\frac{1}{16}-\frac{1}{4}$	$+\frac{1}{4}-1$	+1 -4	+4 -16	+16 -64	+64 mm
17	61	22	2.7-3.8	19	33	9	5	13	21	0
			3.8-4.4	20	36	11	5	15	10	3
			4.4-5.3	11	59	24	1	2	3	0
			Mean	17	43	15	3	9	12	1

SE 45 NE 11 **4544 5995** **Pasture Hill** **Block A**

Surface level +20.0 m
Water not encountered
December 1977

Overburden 1.5 m
Mineral 6.3 m
Bedrock 1.0 m+

LOG

Geological classification	Lithology	Thickness m	Depth m
	Soil	0.4	0.4
Till	Clay, sandy, mottled brown	1.1	1.5
Glacial Sand and Gravel	Sandy Gravel Gravel: fine, subrounded to rounded, equant to tabular, sandstone with limestone and some angular to subrounded chert Sand: fine, rounded to subangular, equant quartz with subrounded to subangular lithic grains	6.3	7.8
Sherwood Sandstone Group	Sandstone, brown	1.0+	8.8

GRADING

Mean for deposit percentages: Fines	Sand	Gravel	Depth below surface (m)	percentages: Fines $-\frac{1}{16}$	Sand $+\frac{1}{16}-\frac{1}{4}$	$+\frac{1}{4}-1$	$+1-4$	Gravel $+4-16$	$+16-64$	$+64$ mm
3	61	36	1.5-2.5	10	69	20	1	0	0	0
			2.5-3.5	4	59	14	14	7	2	0
			3.5-4.5	2	13	21	22	24	18	0
			4.5-5.5	1	7	10	17	30	31	4
			5.5-6.5	2	7	11	25	34	18	3
			6.5-7.8	1	10	17	29	27	14	2
			Mean	3	27	16	18	21	14	1

SE 45 NE 12 **4638 5976** **Thorpe Underwood** **Block A**

Surface level +16.8 m
Water not encountered
October 1977

Waste 11.7 m
Bedrock 0.4 m+

LOG

Geological classification	Lithology	Thickness m	Depth m
	Soil	1.0	1.0
Glacial Lake Deposits	Clays and silt interbedded, laminated in part, with some thin seams of 'clayey' sand	8.5	9.5
Till	Clay, sandy, pebbly, dark brown	2.2	11.7
Sherwood Sandstone Group	Sandstone, brown, weathered to 12.0 m	0.4+	12.1

SE 45 NE 13 **4741 5995** **Pear Tree Cottage Farm** **Block B**

Surface level +14.0 m
Water struck at +4.9 m
October 1977

Waste 11.5 m
Bedrock 2.2 m+

LOG

Geological classification	Lithology	Thickness m	Depth m
	Soil	0.3	0.3
Glacial Lake Deposits	Clay, silty, mottled grey	1.0	1.3
	Clays, silts and sand, laminated and interbedded clays becoming dominant below 3.5 m	7.1	8.4
Till	Clay, sandy, pebbly	0.9	9.3
Fluvio-glacial and Older River Sand and Gravel	Gravel Gravel: coarse and fine, subangular to well-rounded sandstone with limestone and dolerite Sand: coarse, medim and fine, quartz with lithic grains	2.2	11.5
Sherwood Sandstone Group	Sandstone, brown, weathered to 12.8 m	2.2+	13.7

SE 45 NE 14 **4852 5959** **Widdington Manor** **Block B**

Surface level +13.9 m
Water struck at -1.4 m
November 1977

Waste 21.3 m
Bedrock 0.8 m+

LOG

Geological classification	Lithology	Thickness m	Depth m
	Soil	0.9	0.9
Glacial Lake Deposits	Clay, interbedded with sand and silt, laminated in part, brown	12.9	13.8
Till	Clay, sandy, pebbly, brown, with sporadic thin seams of 'very clayey' pebbly sand	3.3	17.1
Fluvio-glacial and Older River Sand and Gravel	Sandy Gravel Gravel: fine and coarse, rounded sandstone and limestone Sand: medium and fine with coarse, quartz with lithic grains	4.2	21.3
Sherwood Sandstone Group	Sandstone, brown, upper surface weathered	0.8+	22.1

SE 45 NE 15 **4910 5964** **Widdington Hall Farm** **Block B**

Surface level +14.0 m
Water struck at +1.5 m
November 1977

Overburden 12.5 m
Mineral 9.0 m
Bedrock 1.3 m+

LOG

Geological classification	Lithology	Thickness m	Depth m
	Made ground	0.3	0.3
	Soil	0.7	1.0
Glacial Lake Deposits	Clay, silty, laminated, brown, with partings of fine-grained sand and silt	11.5	12.5
Fluvio-glacial and Older River Sand and Gravel, ?Till between 12.5 m and 13.5 m	Sandy Gravel Gravel: fine, angular to rounded, tabular and equant, sandstone with limestone and some angular to subangular chert, and rounded ironstone Sand: medium, angular to subrounded, equant quartz, with lithic grains Fines: thin seams of pebbly clay between 12.5 m and 13.5 m	9.0	21.5
Sherwood Sandstone Group	Sandstone, brown	1.3+	22.8

GRADING

Mean for deposit percentages: Fines	Sand	Gravel	Depth below surface (m)	percentages: Fines	Sand			Gravel		
				$-\frac{1}{16}$	$+\frac{1}{16}-\frac{1}{4}$	$+\frac{1}{4}-1$	+1 -4	+4 -16	+16 -64	+64 mm
4	66	30	12.5-13.5	11	61	22	1	1	1	3
			13.5-16.1	3	18	31	15	23	10	0
			16.1-17.3	1	21	40	9	9	17	3
			17.3-18.5	5	29	33	9	12	12	0
			18.5-19.3	3	13	29	13	22	20	0
			19.3-20.4	2	8	29	16	26	19	0
			20.4-21.5	6	13	42	10	14	15	0
			Mean	4	23	32	11	17	12	1

SE 45 NE 16 **4947 5908** **Widdington Wood** **Block B**

Surface level +13.6 m
Water struck at -1.1 m
November 1977

Overburden 14.7 m
Mineral 6.5 m
Bedrock 1.4 m+

LOG

Geological classification	Lithology	Thickness m	Depth m
	Soil	1.1	1.1
Glacial Lake Deposits, ?Sandy Till below 14.7 m	Silt and clay, laminated and interbedded with fine-grained sand	13.6	14.7
Fluvio-glacial and Older River Sand and Gravel	Sandy Gravel Gravel: fine, subangular to rounded, tabular and equant sandstone, with subrounded to rounded, equant limestone and some rounded, tabular ironstone and subangular, equant chert Sand: fine, angular to rounded, equant, quartz with some lithic grains	6.5	21.2
Sherwood Sandstone Group	Sandstone, brown	1.4+	22.6

GRADING

Mean for deposit percentages			Depth below surface (m)	percentages						
Fines	Sand	Gravel		Fines	Sand			Gravel		
				$-\frac{1}{16}$	$+\frac{1}{16}-\frac{1}{4}$	$+\frac{1}{4}-1$	+1 -4	+4 -16	+16 -64	+64 mm
8	67	25	14.7-17.3	15	50	26	3	2	4	0
			17.3-18.7	4	14	19	19	25	19	0
			18.7-19.6	2	17	27	17	27	10	0
			19.6-21.2	3	22	25	19	18	11	2
			Mean	8	31	24	12	14	10	1

SE 45 NE 17 **4547 5888** **Thorpe Hill Farm** **Block A**

Surface level +23.2 m
Water struck at +20.4 m
October 1977

Overburden 2.8 m
Mineral 1.2 m
Waste 2.2 m
Bedrock 2.3 m+

LOG

Geological classification	Lithology	Thickness m	Depth m
	Soil	0.3	0.3
Till	Clay, silty, pebbly, brown, sporadic thin seams of 'clayey' pebbly sand	1.1	1.4
Laminated Clay closely associated with Glacial Deposits	Clay, sandy, laminated, thin seam of 'clayey' sand at 1.7 m, sporadic plant remains	1.4	2.8
Glacial Sand and Gravel	'Very clayey' sand: fine, quartz with lithic grains, laminated	1.2	4.0
Laminated Clay, closely associated with Glacial Deposits	Clay and silt, laminated	1.8	5.8
Glacial Sand and Gravel	'Clayey' gravel Gravel: cobbles, subrounded, sandstone and limestone with shale and some cherts Sand: fine, quartz with some lithic grains	0.4	6.2
Sherwood Sandstone Group	Mudstone, reddish brown	2.3+	8.5

GRADING

Mean for deposit percentages			Depth below surface (m)	percentages						
Fines	Sand	Gravel		Fines	Sand			Gravel		
				$-\frac{1}{16}$	$+\frac{1}{16}-\frac{1}{4}$	$+\frac{1}{4}-1$	+1 -4	+4 -16	+16 -64	+64 mm
29	71	0	2.8-4.0	29	70	1	0	0	0	0

SE 45 NE 18 **4641 5853** **Divel Covert** **Block A**

Surface level +17.5 m
Water struck at +15.2 m
October 1977

Overburden 2.3 m
Mineral 1.5 m
Waste 3.4 m
Mineral 4.7 m
Bedrock 1.4 m+

LOG

Geological classification	Lithology	Thickness m	Depth m
	Soil	0.8	0.8
Glacial Lake Deposits	Clay, mottled grey, sandy in parts, laminated below 1.4 m	1.5	2.3
Sandy Till	**a** 'Very clayey' pebbly sand Gravel: fine, rounded sandstone and subangular to rounded limestone Sand: fine, quartz and lithic grains	1.5	3.8
Till	Clay, sandy, pebbly, brown	3.4	7.2
Sandy Till	**b** 'Very clayey' sand: fine, quartz with lithic grains, with some limestone gravel below 10.6 m	4.7	11.9
Sherwood Sandstone Group	Mudstone, reddish brown, weathered to 12.8 m	1.4+	13.3

GRADING

	Mean for deposit percentages			Depth below surface (m)	percentages						
	Fines	Sand	Gravel		Fines	Sand			Gravel		
					$-\frac{1}{16}$	$+\frac{1}{16}-\frac{1}{4}$	$+\frac{1}{4}-1$	+1 -4	+4 -16	+16 -64	+64 mm
a	23	63	14	2.3-3.8	23	49	11	3	10	4	0
b	22	77	1	7.2-9.4	20	69	11	0	0	0	0
				9.4-10.6	26	62	10	2	0	0	0
				10.6-11.9	21	66	7	3	1	2	0
				Mean	22	66	10	1	0	1	0
a + b	22	74	4	Mean	22	62	10	2	3	1	0

SE 45 NE 19 **4743 5866** **Moss Hill Farm** **Block A**

Surface level +14.0 m
Water struck at -1.5 m
December 1977

Waste 15.5 m
Bedrock 1.0 m+

LOG

Geological classification	Lithology	Thickness m	Depth m
	Made ground	1.6	1.6
Glacial Lake Deposits	Silt, sandy, grey	1.1	2.7
	Clay, dark brown, laminated in part	10.7	13.4
?Till	Silt, sandy, with some fine-grained limestone gravel	2.1	15.5
Sherwood Sandstone Group	Sandstone, brown, upper surface weathered	1.0+	16.5

SE 45 NE 20 **4852 5819** **Tomtit's Nest** **Block B**

Surface level +13.6 m
Water struck at -2.1 m
November 1977

Waste 17.6 m
Bedrock 1.7 m+

LOG

Geological classification	Lithology	Thickness m	Depth m
	Soil	0.7	0.7
Glacial Lake Deposits	Clay, interbedded with silt and sand, laminated in part, sporadic pebbles of sandstone	14.1	14.8
Till	Clay, sandy, pebbly, brown	0.9	15.7
Fluvio-glacial and Older River Sand and Gravel	'Very clayey' pebbly sand Gravel: coarse, fine and cobbles, rounded sandstone Sand: fine, quartz with some lithic grains	1.9	17.6
Sherwood Sandstone Group	Sandstone with mudstone brown, weathered to 19.2	1.7+	19.3

SE 45 NE 21 **4532 5786** **Whixley Field House** **Block A**

Surface level +27.2 m
Water struck at +22.3 m
October 1977

Waste 6.4 m
Bedrock 2.9 m+

LOG

Geological classification	Lithology	Thickness m	Depth m
	Soil	0.5	0.5
Till	Clay, sandy, pebbly, brown	4.1	4.6
	Clay, sandy, pebbly, with seams of 'clayey' sandy gravel and 'clayey' pebbly sand	1.8	6.4
Sherwood Sandstone Group	Sandstone, brown, weathered	2.9+	9.3

SE 45 NE 22 **4643 5733** **Green Hammerton** **Block A**

Surface level +18.1 m
Water struck at + 14.5 m
December 1977

Waste 3.6 m
Bedrock 1.0 m+

LOG

Geological classification	Lithology	Thickness m	Depth m
	Soil	0.6	0.6
Till	Clay, sandy, sporadic pebbles, dark brown	1.8	2.4
	Silt, sandy, pebbly, brown	1.2	3.6
Sherwood Sandstone Group	Sandstone, brown, weathered	1.0+	4.6

SE 45 NE 23 **4746 5794** **Stone's Farm** **Block B**

Surface level +13.2 m
Water struck at -0.9 m
December 1979

Waste 16.5 m
Bedrock 1.1 m+

LOG

Geological classification	Lithology	Thickness m	Depth m
	Soil	0.7	0.7
Glacial Lake Deposits	Clay, interbedded with sand and silt, laminated, sporadic pebbles of limestone and sandstone below 12.1 m	13.4	14.1
?Fluvio-glacial and Older River Sand and Gravel, or ?Glacial Sand and Gravel	'Clayey' sandy gravel Gravel: coarse with fine, subangular to subrounded, tabular and equant, sandstone, limestone and dolerite Sand: fine, quartz with some lithic grains.	2.4	16.5
Sherwood Sandstone Group	Sandstone, weathered to 17.2 m	1.1+	17.6

SE 45 NE 24 **4837 5717** **Pool Bridge Farm** **Block B**

Surface level +13.6 m
Water struck at -1.8 m
December 1977

Waste 15.4 m
Bedrock 1.0 m+

LOG

Geological classification	Lithology	Thickness m	Depth m
	Soil	0.9	0.9
Glacial Lake Deposits, ?Till below 13.0 m	Clay, laminated, with sporadic thin partings of silt or sand, dark brown, pebbles at base	12.9	13.8
Fluvio-glacial and Older River Sand and Gravel	'Very clayey' pebbly sand Gravel: fine, rounded sandstone, equant and tabular, subrounded to rounded limestone Sand: fine, quartz with lithic grains	1.6	15.4
Sherwood Sandstone Group	Sandstone, brown, uppper surface weathered	1.0+	16.4

SE 45 NE 25 **4940 5767** **Monk's Approach** **Block B**

Surface level +14.1 m
Water struck at +8.1 m
November 1977

Waste 5.2 m
Bedrock 4.5 m+

LOG

Geological classification	Lithology	Thickness m	Depth m
	Soil	1.0	1.0
Glacial Lake Deposits	Clay, poorly laminated, sporadic thin partings of silt, grey	3.2	4.2
Till	Clay, sandy, pebbly, brown	1.0	5.2
Sherwood Sandstone Group	Sandstone, brown, weathered	4.5+	9.7

SE 45 NE 26 **4564 5639** **Coney Garth Hill** **Block A**

Surface level +38.4 m
Water level not recorded
December 1977

Overburden 0.8 m
Mineral 1.8 m
Waste 0.3 m
Bedrock 1.7 m+

LOG

Geological classification	Lithology	Thickness m	Depth m
	Soil	0.8	0.8
Sandy Till	'Very Clayey' sand: fine, subangular to subrounded, equant quartz with some lithic grains	1.8	2.6
Till	Clay, pebbly, brown	0.3	2.9
Sherwood Sandstone Group	Sandstone, brown	1.7+	4.6

GRADING

Mean for deposit percentages			Depth below surface (m)	percentages						
Fines	Sand	Gravel		Fines	Sand			Gravel		
				$-\frac{1}{16}$	$+\frac{1}{16}-\frac{1}{4}$	$+\frac{1}{4}-1$	+1 -4	+4 -16	+16 -64	+64 mm
27	73	0	0.8-1.8	20	70	9	1	0	0	0
			1.8-2.6	36	55	7	2	0	0	0
			Mean	27	63	8	2	0	0	0

SE 45 NE 27 **4768 5672** **Moor Lands** **Block B**

Surface level +12.5 m
Water struck at +1.4 m
October 1977

Waste 13.0 m
Bedrock 3.3 m+

LOG

Geological classification	Lithology	Thickness m	Depth m
	Soil	1.4	1.4
Glacial Lake Deposits	Clay, silty, laminated in part, silty parting below 3.5 m, sporadic pebbles below 6.6 m	9.2	10.6
Till	Clay, sandy, pebbly, upper surface composed of cobbles of sandstone	0.5	11.1
Fluvio-glacial and Older River Sand and Gravel	Sandy Gravel Gravel: coarse, angular to rounded, limestone and sandstone Sand: medium, angular quartz with lithic grains	1.9	13.0
Sherwood Sandstone Group	Sandstone, brown, weathered to 13.6 m	3.3+	16.3

SE 45 NE 28 **4845 5632** **Gowlands** **Block B**

Surface level +10.0 m
Water struck at +6.5 m
September 1977

Overburden 3.5 m
Mineral 3.6 m
Bedrock 3.1 m+

LOG

Geological classification	Lithology	Thickness m	Depth m
	Soil	0.4	0.4
Alluvium	Clay, greyish brown, sandy in part	3.1	3.5
	'Clayey' pebbly sand Gravel: coarse, subrounded, sandstone with limestone Sand: fine and medium, quartz with lithic grains	3.6	7.1
Sherwood Sandstone Group	Sandstone, reddish brown, weathered to 9.2 m	3.1+	10.2

GRADING

Mean for deposit percentages			Depth below surface (m)	percentages						
Fines	Sand	Gravel		Fines	Sand			Gravel		
				$-\frac{1}{16}$	$+\frac{1}{16}-\frac{1}{4}$	$+\frac{1}{4}-1$	+1 -4	+4 -16	+16 -64	+64 mm
16	70	14	3.5-5.0	20	37	28	5	5	5	0
			5.0-6.1	10	26	31	7	5	11	10
			6.1-7.1	18	34	37	4	2	5	0
			Mean	16	33	32	5	4	7	3

SE 45 NE 29 **4959 5647** **Finkle Holme** **Block B**

Surface level +14.0 m
Water struck at +3.5 m
September 1977

Overburden 10.5 m
Mineral 3.6 m
Bedrock 2.8 m+

LOG

Geological classification	Lithology	Thickness m	Depth m
	Soil	0.9	0.9
Glacial Lake Deposits	Clay, silty, laminated, dark brown	6.3	7.2
Glacial Sand and Gravel	'Clayey' sandy gravel Gravel: coarse, subrounded, sandstone and limestone Sand: fine, quartz with lithic grains	1.8	9.0
Till	Clay, silty, pebbly, mottled, sandy below 9.5 m	1.5	10.5
Fluvio-glacial and Older River Sand and Gravel	'Clayey' sand: fine, quartz with lithic grains, with some subrounded sandstone below 13.0 m	3.6	14.1
Sherwood Sandstone Group	Sandstone, reddish brown, weathered	2.8+	16.9

GRADING

Mean for deposit percentages			Depth below surface (m)	percentages						
Fines	Sand	Gravel		Fines	Sand			Gravel		
				$-\frac{1}{16}$	$+\frac{1}{16}-\frac{1}{4}$	$+\frac{1}{4}-1$	+1 -4	+4 -16	+16 -64	+64 mm
15	82	3	10.5-12.0	22	74	4	0	0	0	0
			12.0-13.0	9	48	41	2	0	0	0
			13.0-14.1	11	32	40	6	3	8	0
			Mean	15	54	25	3	1	2	0

SE 45 NE 30 **4536 5521** **Westfield** **Block A**

Surface level +28.9 m
Wate struck at +26.9 m
October 1977

Overburden 2.0 m
Mineral 4.0 m
Waste 8.4 m
Bedrock 0.9 m+

LOG

Geological classification	Lithology	Thickness m	Depth m
	Soil	0.3	0.3
Till	Clay, sandy, pebbly, brown, thin seam of 'clayey' sand 0.3 m-0.9 m	1.7	2.0
Sandy Till	'Very clayey' pebbly sand Gravel: fine, coarse and cobbles, rounded, sandstone and limestone with some shale, cherts and volcanics Sand: fine, quartz with lithic grains	4.0	6.0
Till	Clay, sandy, pebbly, brown	8.4	14.4
Sherwood Sandstone Group	Sandstone, reddish brown, weathered	0.9+	15.3

GRADING

Mean for deposit percentages			Depth below surface (m)	percentages						
Fines	Sand	Gravel		Fines	Sand			Gravel		
				$-\frac{1}{16}$	$+\frac{1}{16}-\frac{1}{4}$	$+\frac{1}{4}-1$	+1 -4	+4 -16	+16 -64	+64 mm
29	54	17	2.0-3.0	27	52	10	2	6	3	0
			3.0-4.0	30	29	10	2	7	2	20
			4.0-5.0	36	45	7	3	8	1	0
			5.0-6.0	24	44	7	3	9	13	0
			Mean	29	43	9	2	7	5	5

SE 45 NE 31 **4648 5503** **The Park** **Block D**

Surface level +16.8 m
Water struck at +15.1 m
November 1977

Overburden 0.2 m
Mineral 5.0 m
Waste 2.3 m
Mineral 1.0 m
Waste 3.5 m
Mineral 3.1 m
Bedrock 0.9 m+

LOG

Geological classification	Lithology	Thickness m	Depth m
	Soil	0.2	0.2
Glacial Lake Deposit	**a** 'Clayey' sand: fine, subangular to subrounded, equant quartz with lithic grains	5.0	5.2
	Clay, laminated, brown, with partings of silt and fine-grained sand	2.1	7.3
Till	Clay, sandy, pebbly, dark brown	0.2	7.5
Sandy Till	**b** 'Very clayey' pebbly sand Gravel: fine and coarse, sandstone and limestone Sand: fine, subangular to subrounded, equant, quartz with lithic grains	1.0	8.5
Till	Clay, sandy, pebbly, brown	3.5	12.0
Fluvio-glacial and Older River Sand and Gravel	**c** Sandy Gravel Gravel: coarse, angular to subrounded, equant sandstone, with some subrounded to rounded, equant quartz and limestone Sand: medium, angular to subrounded, equant quartz with lithic grains	3.1	15.1
Sherwood Sandstone Group	Mudstone, brown	0.9+	16.0

GRADING

	Mean for deposit percentages			Depth below surface (m)	percentages						
	Fines	Sand	Gravel		Fines	Sand			Gravel		
					$-\frac{1}{16}$	$+\frac{1}{16}-\frac{1}{4}$	$+\frac{1}{4}-1$	+1 -4	+4 -16	+16 -64	+64 mm
a	13	86	1	0.2-1.7	17	43	31	6	3	0	0
				1.7-5.2	12	76	11	1	0	0	0
				Mean	13	66	17	3	1	0	0
b	27	62	11	7.5-8.5	27	47	11	4	5	6	0
c	4	49	47	12.0-13.3	4	8	20	14	20	34	0
				13.3-14.0	6	13	20	11	19	25	6
				14.0-15.1	3	14	34	13	14	19	3
				Mean	4	11	25	13	18	26	3
a+b+c	12	70	18	Mean	12	45	19	6	7	10	1

SE 45 NE 32 **4813 5532** **Wilstrop Hall** **Block B**

Surface level +11.0 m
Water struck at +0.4 m
December 1977

Overburden 1.5 m
Mineral 1.0 m
Waste 1.5 m
Mineral 3.8 m
Waste 1.1 m
Mineral 1.3 m
Waste 0.4 m
Mineral 2.2 m
Bedrock 1.0 m+

LOG

Geological classification	Lithology	Thickness m	Depth m
	Soil	0.5	0.5
Alluvium	Clay, sandy, mottled, brown	1.0	1.5
	a 'Very clayey' pebbly sand: fine, quartz with lithic grains and some gravel	1.0	2.5
	Silt, sandy, brown	0.9	3.4
	Clay, sandy, brown	0.6	4.0
	b Pebbly sand Gravel: coarse and fine, sandstone and limestone Sand: medium, subangular to subrounded, quartz with lithic grains	3.8	7.8
Glacial Lake Deposits	Clay, sandy, poorly laminated, brown	1.1	8.9
	c 'Clayey' sand: fine, subangular to subrounded quartz with lithic grains	1.3	10.2
?Till	Clay, laminated with sporadic pebbles, brown	0.4	10.6
Fluvio-glacial and Older River Sand and Grave	**d** Gravel Gravel: caorse, subrounded to rounded, equant sandstone with limestone Sand: fine, subangular to subrounded, equant, quartz with lithic grains	2.2	12.8
Sherwood Sandstone Group	Sandstone, brown	1.0+	13.8

GRADING

	Mean for deposit percentages			Depth below surface (m)	percentages						
	Fines	Sand	Gravel		Fines	Sand			Gravel		
					$-\frac{1}{16}$	$+\frac{1}{16}-\frac{1}{4}$	$+\frac{1}{4}-1$	+1 -4	+4 -16	+16 -64	+64 mm
a	25	59	16	1.5-2.5	25	46	11	2	7	9	0
b	6	89	5	4.0-5.0	1	39	46	3	3	8	0
				5.0-6.0	11	39	45	3	2	0	0
				6.0-7.0	6	27	60	3	2	2	0
				7.0-7.8	8	21	65	2	1	3	0
				Mean	6	32	54	3	2	3	0
c	15	85	0	8.9-10.2	15	74	10	1	0	0	0
d	5	39	56	10.6-11.6	5	16	8	3	16	50	2
				11.6-12.8	5	21	19	8	16	27	4
				Mean	5	19	14	6	16	37	3
a+b+ c+d	9	72	19	Mean	9	37	31	4	6	12	1

SE 45 NE 33 **4943 5536** **Rose Cottage Farm** **Block B**

Surface level +13.6 m
Water struck at +4.6 m
August 1977

Overburden 4.9 m
Mineral 9.4 m
Bedrock 0.6 m+

LOG

Geological classification	Lithology	Thickness m	Depth m
	Soil	0.4	0.4
Glacial Lake Deposits	Clay, sandy, brown, laminations of grey silt below 2.8 m and of fine-grained sand below 3.8 m	4.5	4.9
?Sandy Till	**a** 'Very clayey' pebbly sand Gravel: coarse and fine, subrounded to rounded sandstone with mudstone Sand: fine, quartz with some lithic grains	6.8	11.7
Fluvio-glacial and Older River Sand and Gravel	b Gravel Gravel: coarse, subrounded to rounded, sandstone with some calcareous mudstone Sand: fine and medium, quartz with some lithic grains	2.6	14.3
Sherwood Sandstone Group	Sandstone, reddish brown	0.6+	14.9

GRADING

	Mean for deposit percentages			Depth below surface (m)	percentages						
	Fines	Sand	Gravel		Fines	Sand			Gravel		
					$-\frac{1}{16}$	$+\frac{1}{16}-\frac{1}{4}$	$+\frac{1}{4}-1$	+1 -4	+4 -16	+16 -64	+64 mm
a	28	66	6	4.9-6.0	33	33	11	2	7	14	0
				6.0-7.0	17	49	26	3	3	2	0
				7.0-8.0	36	40	11	3	2	8	0
				8.0-9.1	31	64	5	0	0	0	0
				9.1-10.7	22	62	15	1	0	0	0
				10.7-11.7	28	48	13	4	4	3	0
				Mean	28	50	14	2	2	4	0
b	8	45	47	11.7-12.7	8	17	11	9	14	41	0
				12.7-14.3	8	19	19	12	10	32	0
				Mean	8	18	16	11	12	35	0
a+b	22	60	18	Mean	22	42	14	4	5	13	

SE 45 SW 67 **4062 5485** **New Cut** **Block C**

Surface level +28.6 m
Water struck at +19.4 m
January 1978

Waste 14.7 m
Bedrock 0.5 m+

LOG

Geological classification	Lithology	Thickness m	Depth m
	Soil, sandy and silty clay	0.5	0.5
Glacial Lake Deposits	Sand: medium, subrounded quartz with lithic grains, sporadic thin seams of clay	0.5	1.0
Till	Clay, sandy, pebbly, brown	8.0	9.0
Clay associated with Fluvio-glacial and Older River Sand and Gravel	Clay and silt, laminated with thin seams of sand, brown	2.2	11.2
Fluvio-glacial and Older River Sand and Gravel	Sandy Gravel Gravel: coarse, subangular, equant, limestone and rounded, equant sandstone with some angular chert Sand: medium, subangular to rounded, equant, quartz with lithic grains	3.5	14.7
Sherwood Sandstone Group	Sandstone	0.5+	15.2

SE 45 SW 68 **4187 5462** **New Forest Farm** **Block C**

Surface level +26.2 m
Water struck at +23.9 m
January 1978

Overburden 2.3 m
Mineral 1.1 m
Waste 8.6 m
Bedrock 0.8 m+

LOG

Geological classification	Lithology	Thickness m	Depth m
	Made ground	0.6	0.6
Glacial Lake Deposits	Clay, sandy, poorly laminated in part, mottled greyish brown	0.9	1.5
Till	Clay, pebbly, brown	0.8	2.3
Sandy Till	'Very clayey' pebbly sand Gravel: coarse, angular to rounded, equant, sandstone and limestone Sand: fine, angular to subrounded, equant, quartz with some lithic grains.	1.1	3.4
Till	Clay, sandy, pebbly, brown	8.0	11.4
?Clay associated with Fluvio-glacial and Older River Sand and Gravel	Clay, laminated with sporadic pebbles	0.6	12.0
Sherwood Sandstone Group	Sandstone	0.8+	12.8

GRADING

Mean for deposit percentages			Depth below surface (m)	percentages						
Fines	Sand	Gravel		Fines	Sand			Gravel		
				$-\frac{1}{16}$	$+\frac{1}{16}-\frac{1}{4}$	$+\frac{1}{4}-1$	+1 -4	+4 -16	+16 -64	+64 mm
22	60	18	2.3-3.4	22	43	11	6	6	8	4

SE 45 SW 69 **4300 5436** **Nursery Hill Cottage** **Block C**

Surface level +45.6 m
Water struck at +30.3 m
January 1978

Overburden 0.5 m
Mineral 3.2 m
Waste 0.3 m
Mineral 15.1 m
Waste 5.9 m+

LOG

Geological classification	Lithology	Thickness m	Depth m
	Soil	0.5	0.5
Glacial Sand and Gravel	**a** 'Clayey' sandy gravel Gravel: fine and coarse, angular to subrounded, equant sandstone and limestone Sand: fine, subangular to rounded, equant, quartz with with lithic grains	3.2	3.7
Till	Clay, sandy, sporadic thin seams of sand and scattered pebbles	0.3	4.0
Glacial Sand and Gravel	**b** 'Clayey' sand: fine, subangular to subrounded, equant, quartz with some lithics; sporadic thin seams of laminated clays and silts	15.1	19.1
Till	Clay, sandy, pebbly, brown	5.9+	25.0

GRADING

	Mean for deposit percentages			Depth below surface (m)	percentages						
	Fines	Sand	Gravel		Fines	Sand			Gravel		
					$-\frac{1}{16}$	$+\frac{1}{16}-\frac{1}{4}$	$+\frac{1}{4}-1$	+1 -4	+4 -16	+16 -64	+64 mm
a	10	46	44	0.5-1.8	11	32	13	7	14	20	3
				1.8-2.8	6	13	12	14	28	27	0
				2.8-3.7	14	30	8	11	22	15	0
				Mean	10	25	11	10	21	21	2
b	16	81	3	4.0-5.0	23	55	10	3	7	2	0
				5.0-6.0	10	75	15	0	0	0	0
				6.0-7.0	7	77	15	1	0	0	0
				7.0-8.1	12	75	12	1	0	0	C
				8.1-9.2	24	64	12	0	0	0	0
				9.2-10.2	12	80	7	1	0	0	0
				10.2-11.2	28	62	8	1	1	0	0
				11.2-12.2	25	68	5	1	1	0	0
				12.2-13.2	16	73	6	2	3	0	0
				13.2-14.4	18	69	10	2	1	0	0
				14.4-15.3	10	68	14	1	1	6	0
				15.3-16.4	8	73	18	1	0	0	0
				16.4-19.1	17	71	5	1	1	0	5
				Mean	16	70	10	1	1	1	1
a + b	15	75	10	Mean	15	62	10	3	5	4	1

SE 45 SW 70 **4432 5463** **Hamley Hill** **Block D**

Surface level +23.7 m
Water struck at +12.7 m
December 1977

Overburden 1.1 m
Mineral 11.0 m
Waste 1.8 m
Mineral 8.0 m
Bedrock 1.6 m+

LOG

Geological classification	Lithology	Thickness m	Depth m
	Soil	0.5	0.5
Till	Clay, sandy, brown	0.6	1.1
Sandy Till	**a** 'Very clayey' pebbly sand Gravel: coarse and fine, subangular to subrounded, equant, sandstone and limestone Sand: fine, subangular to rounded, equant, quartz with lithic grains	11.0	12.1
Till	Clay, sandy, pebbly, brown	0.6	12.7
?Glacial Sand and Gravel	Clayey sand: fine, subangular to rounded, quartz with lithic grains	0.9	13.6
Clay associated with Fluvio-glacial and older River Sand and Gravel	Clay, laminated, greyish brown	0.3	13.9
Fluvio-glacial and Older River Sand and Gravel	**b** 'Clayey' sand: fine, subangular to rounded equant, quartz with lithic grains	5.0	18.9
	c Gravel Gravel: coarse, subrounded to rounded, equant sandstone with limestone Sand: medium, subangular to rounded, quartz with lithic grains	3.0	21.9
Sherwood Sandstone Group	Sandstone	1.6+	23.5

GRADING

	Mean for deposit percentages			Depth below surface (m)	percentages						
	Fines	Sand	Gravel		Fines	Sand			Gravel		
					$-\frac{1}{16}$	$+\frac{1}{16}-\frac{1}{4}$	$+\frac{1}{4}-1$	+1 -4	+4 -16	+16 -64	+64 mm
a	34	57	9	1.1-2.1	28	53	13	6	0	0	0
				2.1-3.1	31	50	9	2	3	5	0
				3.1-4.1	35	46	11	8	0	0	0
				4.1-5.1	33	42	9	3	4	9	0
				5.1-6.1	35	42	7	3	4	9	0
				6.1-7.1	39	43	10	4	4	0	0
				7.1-8.1	34	41	9	4	7	5	0
				8.1-9.1	37	44	8	4	4	3	0
				9.1-10.1	34	41	8	3	4	10	0
				10.1-11.1	34	40	7	2	4	7	6
				11.1-12.1	30	41	7	3	6	13	0
				Mean	34	44	9	4	4	5	0
b	19	78	3	13.9-14.9	29	55	7	4	5	0	0
				14.9-15.9	17	71	5	1	1	5	0
				15.9-16.9	11	73	15	1	0	0	0
				16.9-17.9	20	72	8	0	0	0	0
				17.9-18.9	17	69	10	0	1	3	0
				Mean	19	68	9	1	1	2	0
c	3	42	55	18.9-19.9	1	7	14	10	32	36	0
				19.9-20.7	3	12	26	14	21	24	0
				20.7-21.9	4	12	21	12	15	33	3
				Mean	3	10	20	12	22	32	1
a+b+c	25	60	15	Mean	25	45	11	4	6	9	0

SE 45 SW 71 **4003 5315** **Black Stones** **Block C**

Surface level +28.2 m
Water struck at +19.4 m
January 1978

Overburden 8.8 m
Mineral 3.1 m
Waste 1.2 m+

LOG

Geological classification	Lithology	Thickness m	Depth m
	Soil	0.8	0.8
River Terrace Deposits, undifferentiated	Clay, sandy, carbonaceous lenses, brown	0.4	1.2
	Silt, sandy, pebbly, brown	1.1	2.3
	Gravel Gravel: coarse, rounded to well-rounded, tabular, sandstone and limestone with some chert Sand: medium, angular, quartz with lithic grains	0.9	3.2
Till	Clay, sandy, pebbly, brown	5.6	8.8
Fluvio-glacial and Older River Sand and Gravel?	Sandy Gravel Gravel: fine, subangular to subrounded, equant sandstone with subrounded to rounded quartz and some ironstone and limestone Sand: medium quartz with lithic grains Fines: sporadic thin seams of silty clay	3.1	11.9
Clay associated with Fluvio-glacial and Older River Sand and Gravel	Clay, silty, brown	1.2+	13.1

GRADING

Mean for deposit percentages			Depth below surface (m)	percentages						
Fines	Sand	Gravel		Fines	Sand			Gravel		
				$-\frac{1}{16}$	$+\frac{1}{16}-\frac{1}{4}$	$+\frac{1}{4}-1$	+1 -4	+4 -16	+16 -64	+64 mm
6	68	26	8.8-10.5	8	17	41	24	10	0	0
			10.5-11.9	4	6	24	23	21	19	3
			Mean	6	12	33	23	15	9	2

SE 45 SW 72 **4040 5383** **Ribston Park** **Block C**

Surface level +27.0 m
Water struck at +24.8 m
December 1977

Overburden 0.9 m
Mineral 1.9 m
Waste 8.3 m
Bedrock 1.2 m+

LOG

Geological classification	Lithology	Thickness m	Depth m
	Soil	0.9	0.9
Alluvium	Gravel Gravel: coarse, rounded to angular, sandstone and limestone Sand: medium, angular, quartz with lithic grains	1.9	2.8
Till	Clay, silty, pebbly, brown, thin seam of laminated fine-grained sand and silt at 8.7 m	6.5	9.3
Fluvio-glacial and Older River Sand and Gravel	Gravel Gravel: fine, rounded to well-rounded, sandstone and limestone with igneous rocks and chert Sand: coarse, subangular to subrounded, quartz with lithic grains	1.8	11.1
Sherwood Sandstone Group	Sandstone, reddish brown	1.2+	12.3

GRADING

Mean for deposit percentages			Depth below surface (m)	percentages						
Fines	Sand	Gravel		Fines	Sand			Gravel		
				$-\frac{1}{16}$	$+\frac{1}{16}-\frac{1}{4}$	$+\frac{1}{4}-1$	+1 -4	+4 -16	+16 -64	+64 mm
9	45	46	0.9-2.8	9	11	20	14	19	27	0

SE 45 SW 73 **4164 5322** **Walshford Bridge** **Block C**

Surface level +25.3 m
Water struck at +14.8 m
November 1977

Overburden 11.8 m
Mineral 4.7 m
Bedrock 1.7 m+

LOG

Geological classification	Lithology	Thickness m	Depth m
Glacial Lake Deposits	Soil	0.3	0.3
	Clay, sandy and silty with pebbly clay layers	1.6	1.9
Till	Clay, sandy, pebbly, brown	9.9	11.8
Fluvio-glacial and Older River Sand and Gravel	Sand: medium, angular, equant, quartz with lithic grains	4.7	16.5
Sherwood Sandstone Group	Sandstone	1.7+	18.2

GRADING

Mean for deposit percentages			Depth below surface (m)	percentages						
Fines	Sand	Gravel		Fines	Sand			Gravel		
				$-\frac{1}{16}$	$+\frac{1}{16}-\frac{1}{4}$	$+\frac{1}{4}-1$	+1 -4	+4 -16	+16 -64	+64 mm
10	86	4	11.8-13.4	11	29	56	3	1	0	0
			13.4-14.8	11	25	51	9	3	1	0
			14.8-16.5	7	19	53	13	5	3	
			Mean	10	24	53	9	3	1	0

SE 45 SW 74 **4448 5367** **Cattal Bridge** **Block C**

Surface level +19.8 m
Water struck at +10.4 m
December 1977

Overburden 0.8 m
Mineral 1.0 mm
Waste 13.1 m
Bedrock 1.1 m+

LOG

Geological classification	Lithology	Thickness m	Depth m
	Soil	0.8	0.8
River Terrace Deposits, undifferentiated	'Clayey' pebbly sand Gravel: fine, subangular to rounded, sandstone with limestone Sand: medium, subangular to rounded, quartz with lithic grains	1.0	1.8
Glacial Lake Deposits	Clay, laminated, with partings of sand and silt, brown	3.7	5.5
?Till	Clay, sandy and silty with pebbles and sand layers, poorly laminated in part	6.5	12.0
Fluvio-glacial and Older River Sand and Gravel	Sandy Gravel Gravel: coarse, subrounded to rounded, sandstone with limestone Sand: medium, subrounded to rounded, quartz with lithic grains	2.9	14.9
Sherwood Sandstone Group	Sandstone, brown	1.1+	16.0

GRADING

Mean for deposit percentages			Depth below surface (m)	percentages						
Fines	Sand	Gravel		Fines	Sand			Gravel		
				$-\frac{1}{16}$	$+\frac{1}{16}-\frac{1}{4}$	$+\frac{1}{4}-1$	+1 -4	+4 -16	+16 -64	+64 mm
18	67	15	0.8-1.8	18	17	39	11	11	4	0

SE 45 SW 75 **4085 5264** **Ruddings Farm** **Block C**

Surface level +24.1 m
Water struck at +14.7 m
December 1977

Waste 10.4 m
Bedrock 2.3 m+

LOG

Geological classification	Lithology	Thickness m	Depth m
	Soil	1.2	1.2
Till	Clay, sandy, pebbly, brown, silty in part, laminated and silty below 8.5 m	8.2	9.4
Fluvio-glacial and Older River Sand and Gravel	Gravel Gravel: fine, sandstone and limestone Sand: medium, quartz and lithic grains	1.0	10.4
Sherwood Sandstone Group	Sandstone, brown, weathered	2.3+	12.7

SE 45 SW 76 4157 5217 Hall Garth **Block C**

Surface level +20.8 m
Water struck at +18.5 m
December 1977

Overburden 0.8 m
Mineral 1.4 m
Waste 6.4 m
Bedrock 1.0 m+

LOG

Geological classification	Lithology	Thickness m	Depth m
	Soil	0.8	0.8
Alluvium	'Very clayey' pebbly sand Gravel: coarse and fine, rounded to well-rounded, bladed and equant, sandstone with subangular limestone and some angular chert Sand: medium and fine, quartz with lithic grains Fines: thin seam of sandy clay	1.4	2.2
Till	Clay, sandy, pebbly, brown	4.9	7.1
Clay associated with Fluvio-glacial and Older River Sand and Gravel	Clay and silt, laminated, with thin seams of pebbly sand	1.5	8.6
Sherwood Sandstone Group	Sandstone, brown	1.0+	9.6

GRADING

Mean for deposit percentages			Depth below surface (m)	percentages						
Fines	Sand	Gravel		Fines	Sand			Gravel		
				$-\frac{1}{16}$	$+\frac{1}{16}-\frac{1}{4}$	$+\frac{1}{4}-1$	+1 -4	+4 -16	+16 -64	+64 mm
24	62	14	0.8-1.5	22	39	37	1	1	0	0
			1.5-1.6	100	0	0	0	0	0	0
			1.6-2.2	14	21	28	7	12	18	0
			Mean	24	28	30	4	6	8	0

SE 45 SW 77 **4256 5266** **Cowthorpe Hall Farm** **Block C**

Surface level +24.8 m
Water struck at +17.8 m
December 1977

Overburden 0.5 m
Mineral 1.2 m
Waste 9.5 m
Mineral 5.5 m
Bedrock 1.6 m+

LOG

Geological classification	Lithology	Thickness m	Depth m
	Soil	0.5	0.5
Glacial Lake Deposits	**a** 'Very clayey' gravel Gravel: fine, subangular to subrounded, sandstone with limestone Sand: fine, subangular to rounded, quartz with lithic grains	1.2	1.7
Till	Clay, sandy, pebbly, brown	4.7	6.4
Clay associated with Fluvio-glacial and Older River Sand and Gravel	Clay, laminated, brown, with seams of 'clayey' sand	4.8	11.2
Fluvio-glacial and Older River Sand and Gravel	**b** 'Clayey' sandy gravel Gravel: coarse, subangular to rounded, sandstone and limestone with some quartz and mudstone Sand: fine, quartz with lithic grains	5.5	16.7
Sherwood Sandstone Group	Sandstone, brown	1.6+	18.3

GRADING

	Mean for deposit percentages			Depth below surface (m)	percentages						
	Fines	Sand	Gravel		Fines	Sand			Gravel		
					$-\frac{1}{16}$	$+\frac{1}{16}-\frac{1}{4}$	$+\frac{1}{4}-1$	+1 -4	+4 -16	+16 -64	+64 mm
a	20	38	42	0.5-1.7	20	19	10	9	17	13	12
b	15	60	25	11.2-12.8	25	53	19	2	1	0	0
				12.8-14.2	20	28	21	7	8	16	0
				14.2-16.7	5	14	26	14	10	24	7
				Mean	15	29	23	8	7	17	3
a + b	16	56	28	Mean	16	27	21	8	9	14	5

SE 45 SW 78 **4473 5259** **Moor Side** **Block D**

Surface level +20.0 m
Water struck at +13.5 m
December 1977

Overburden 10.4 m
Mineral 8.8 m
Bedrock 2.3 m+

LOG

Geological classification	Lithology	Thickness m	Depth m
	Made Ground	0.3	0.3
Glacial Lake Deposits	Clay, sandy, pebbly	0.8	1.1
	Clay, sandy, poorly laminated in part, brown	1.2	2.3
Till	Clay, sandy, pebbly, brown	6.8	9.1
Clay associated with Fluvio-glacial and Older River Sand and Gravel	Silt and clay, laminated, with parting of fine-grained sand	1.3	10.4
Fluvio-glacial and Older River Sand and Gravel	**a** Sand: medium, subangular to rounded, quartz with lithic grains	3.7	14.1
	b Sandy Gravel Gravel: fine, subrounded to well-rounded, equant and tabular, sandstone and limestone with some quartz and chert Sand: as above	5.1	19.2
Sherwood Sandstone Group	Sandstone	2.3+	21.5

GRADING

	Mean for deposit percentages			Depth below surface (m)	percentages						
	Fines	Sand	Gravel		Fines	Sand			Gravel		
					$-\frac{1}{16}$	$+\frac{1}{16}-\frac{1}{4}$	$+\frac{1}{4}-1$	+1 -4	+4 -16	+16 -64	+64 mm
a	5	95	0	10.4-11.7	8	25	53	14	0	0	0
				11.7-12.9	5	12	79	4	0	0	0
				12.9-14.1	3	5	76	16	0	0	0
				Mean	5	14	69	12	0	0	0
b	3	55	42	14.1-15.5	3	6	36	15	31	9	0
				15.5-17.0	2	4	25	14	34	21	0
				17.0-18.2	1	8	40	16	21	14	0
				18.2-19.2	4	9	29	22	24	12	0
				Mean	3	6	33	16	28	14	0
a + b	4	72	24	Mean	4	10	48	14	16	8	0

SE 45 SW 79 **4129 5156** **Deighton Grange** **Block C**

Surface level +25.2 m
Water struck at +14.1 m
August 1977

Overburden 0.4 m
Mineral 1.7 m
Waste 13.3 m
Bedrock 0.1 m+

LOG

Geological classification	Lithology	Thickness m	Depth m
	Soil	0.4	0.4
Sandy Till	'Clayey' pebbly sand Gravel: fine and coarse, subangular to subrounded, sandstone with some limestone and chert Sand: medium, subangular to rounded, quartz with lithic grains	1.7	2.1
Till	Clay, sandy, pebbly, grey becoming brown at 3.8 m, thin seam of laminated sand at 11.1 m	13.3	15.4
Upper Marl	Marl, reddish brown	0.1+	15.5

GRADING

Mean for deposit percentages			Depth below surface (m)	percentages						
Fines	Sand	Gravel		Fines	Sand			Gravel		
				$-\frac{1}{16}$	$+\frac{1}{16}-\frac{1}{4}$	$+\frac{1}{4}-1$	+1 -4	+4 -16	+16 -64	+64 mm
17	78	5	0.4-1.4	16	20	45	13	4	2	0
			1.4-2.1	18	30	36	11	3	2	0
			Mean	17	24	42	12	3	2	0

SE 45 SW 80 **4214 5142** **Oates Wood** **Block C**

Surface level +27.2 m
Water struck at +13.8 m
November 1977

Waste 16.5 m
Bedrock 1.5 m+

LOG

Geological classification	Lithology	Thickness m	Depth m
	Soil	1.3	1.3
Sandy Till	'Clayey' pebbly sand Gravel: fine, sandstone and limestone Sand: fine, quartz and lithic grains	0.8	2.1
Till	Clay, sandy, pebbly, brown, very sandy at 12.8 m	13.3	15.4
Fluvio-glacial and Older River Sand and Gravel	Sandy Gravel Gravel: coarse, sandstone with limestone Sand: medium, quartz with lithic grains	1.1	16.5
Sherwood Sandstone Group	Sandstone and mudstone, brown, weathered	1.5+	18.0

SE 45 SW 81 **4258 5175** **Oates Wood** **Block C**

Surface level +25.9 m
Water struck at +14.4 m
December 1977

Overburden 11.5 m
Mineral 4.9 m
Bedrock 1.9 m+

LOG

Geological classification	Lithology	Thickness m	Depth m
Alluvium	Soil	0.3	0.3
Till	Clay, sandy, pebbly, brown	11.2	11.5
Fluvio-glacial and Older River Sand and Gravel	Gravel Gravel: coarse and fine, subrounded to rounded, sandstone with limestone quartz and chert Sand: medium, angular quartz with rounded lithic grains	4.9	16.4
Sherwood Sandstone Group	Sandstone, reddish brown, weathered	1.9+	18.3

GRADING

Mean for deposit percentages			Depth below surface (m)	percentages						
Fines	Sand	Gravel		Fines	Sand			Gravel		
				$-\frac{1}{16}$	$+\frac{1}{16}-\frac{1}{4}$	$+\frac{1}{4}-1$	+1 -4	+4 -16	+16 -64	+64 mm
2	48	50	11.5-12.8	1	4	26	14	24	31	0
			12.8-14.2	2	6	18	16	31	25	2
			14.2-15.4	5	7	40	16	19	13	0
			15.4-16.4	2	10	24	15	19	30	0
			Mean	2	6	27	15	24	25	1

SE 45 SW 82 **4380 5119** **Lingcroft Plantation** **Block C**

Surface level +25.9 m
Water struck at +18.4 m
December 1977

Waste 19.3 m+

LOG

Geological classification	Lithology	Thickness m	Depth m
	Soil	0.4	0.4
Till	Clay, sandy, pebbly, brown	13.6	14.0
Clay associated with Fluvio-glacial and Older River Sand and Gravel	Silt, sandy in part, laminated, with some seams of silty clay	2.2	16.2
Fluvio-glacial and Older River Sand and Gravel	Sand, coarse, angular quartz with some lithic grains	0.5	16.7
Clay associated with Fluvio-glacial and Older River Sand and Gravel	Clay, silty, greyish brown, pebbly to 18.2, laminated in part	2.6+	19.3

SE 45 SW 83 **4465 5169** **Lingcroft** **Block D**

Surface level +23.5 m
Water struck at +12.2 m
September 1977

Overburden 15.8 m
Mineral 6.0 m
Bedrock 0.5 m+

LOG

Geological classification	Lithology	Thickness m	Depth m
	Soil	0.4	0.4
Till	Clay, sandy, pebbly, brown	10.9	11.3
Laminated Clay closely associated with Glacial Deposits	Clay, laminated, brown, with partings of silt and sand	1.9	13.2
Sandy Till	'Very clayey' pebbly sand Gravel: fine, limestone and sandstone Sand: fine, quartz with lithic grains	1.5	14.7
Till	Clay, silty, pebbly, brown	1.1	15.8
Fluvio-glacial and Older River Sand and Gravel	Sandy Gravel Gravel: coarse and fine, rounded to well-rounded, limestone and sandstones with angular to rounded quartz and chert Sand: medium, angular quartz with rounded lithic grains	6.0	21.8
Sherwood Sandstone Group	Sandstone, reddish brown	0.5+	22.3

GRADING

Mean for deposit percentages			Depth below surface (m)	percentages						
Fines	Sand	Gravel		Fines	Sand			Gravel		
				$-\frac{1}{16}$	$+\frac{1}{16}-\frac{1}{4}$	$+\frac{1}{4}-1$	+1 -4	+4 -16	+16 -64	+64 mm
7	58	35	15-16-7	15	17	54	10	4	0	0
			16.7-18.2	5	11	33	16	19	16	0
			18.2-19.5	9	1	17	21	27	25	0
			19.5-21.8	3	13	27	19	16	22	0
			Mean	7	11	30	17	17	18	0

SE 45 SW 84 **4098 5067** **Loshpot Lane** **Block C**

Surface level +25.7 m
Water struck at +18.9 m
November 1977

Overburden 0.3 m
Mineral 4.3 m
Waste 4.4 m
Bedrock 1.0 m+

LOG

Geological classification	Lithology	Thickness m	Depth m
	Soil	0.3	0.3
Fluvio-glacial Terrace Deposits, undifferentiated	'Clayey' pebbly sand Gravel: fine, subangular to subrounded, sandstone with some rounded limestone, angular chert and quartz Sand: medium, subrounded to rounded, quartz with some lithic grains	4.3	4.6
Till	Clay, sandy, pebbly, grey becoming brown at 5.7 m, thin seam of laminated silt at 6.9 m	4.4	9.0
Upper Magnesian Limestone	Limestone	1.0+	10.0

GRADING

Mean for deposit percentages			Depth below surface (m)	percentages						
Fines	Sand	Gravel		Fines	Sand			Gravel		
				$-\frac{1}{16}$	$+\frac{1}{16}-\frac{1}{4}$	$+\frac{1}{4}-1$	+1 -4	+4 -16	+16 -64	+64 mm
16	68	16	0.3-1.4	11	20	49	11	9	0	0
			1.4-2.9	14	23	43	11	6	3	0
			2.9-4.6	21	22	20	11	13	13	0
			Mean	16	22	35	11	10	6	0

SE 45 SW 85 **4169 5030** **Sugden Wood** **Block C**

Surface level +23.3 m
Water not encountered
November 1977

Waste 10.2 m
Bedrock 1.4 m+

LOG

Geological classification	Lithology	Thickness m	Depth m
Glacial Lake Deposits	Soil	0.3	0.3
	Clay, very sandy, pebbly, brown	0.8	1.1
	'Clayey' sandy gravel Gravel: fine, subangular sandstone with limestone Sand: medium, quartz with lithic grains	0.7	1.8
Till	Clay, sandy, pebbly, brown	8.4	10.2
Upper Marl	Marl, brown to 11.1 m, then grey	1.4+	11.6

SE 45 SW 86 **4247 5007** **Ingmanthorpe Hall Farm** **Block C**

Surface level +34.5 m
Water not encountered
November 1977

Waste 9.1 m
Bedrock 1.3 m+

LOG

Geological classification	Lithology	Thickness m	Depth m
	Soil	0.3	0.3
Till	Clay, sandy, pebbly, brown, sporadic seams of 'clayey' sand	8.8	9.1
Upper Magnesian Limestone	Marl with limestone, pale yellow	1.3+	10.4

SE 45 SW 87 **4276 5082** **Ingmanthorpe** **Block C**

Surface level +27.1 m
Water struck at +9.6 m
November 1977

Waste 19.1 m
Bedrock 0.5 m+

LOG

Geological classification	Lithology	Thickness m	Depth m
	Soil	1.1	1.1
Till	Clay, sandy, pebbly, brown, thin seams of 'clayey' pebbly sand	12.1	13.2
Glacial Sand and Gravel	Sandy Gravel Gravel: fine, subrounded to well-rounded sandstone and limestone Sand: fine, angular, quartz with lithic grains	3.5	16.7
Till	Clay, sandy, pebbly, brown	0.8	17.5
Fluvio-glacial and Older River Sand and Gravel	Gravel Gravel: fine, subrounded to rounded sandstone, chert and limestone Sand: medium, angular, quartz with lithic grain	1.6	19.1
Upper Marl	Marl, calcareous, sandy, brown	0.5+	19.6

SE 45 SW 88 **4358 5027** **Ingmanthorpe Hall Farm** **Block D**

Surface level +25.4 m
Water struck at +13.9 m
November 1977

Overburden 11.5 m
Mineral 4.2 m
Waste 2.8 m
Bedrock 0.5 m+

LOG

Geological classification	Lithology	Thickness m	Depth m
	Soil	1.0	1.0
Till	Clay, sandy, pebbly, brown	10.0	11.0
Clay associated with Fluvio-glacial and Older River Sand and Gravel	Silt and clay, laminated, greyish brown with some fine-grained sand partings	0.5	11.5
Fluvio-glacial and Older River Sand and Gravel	Sandy Gravel Gravel: fine, angular to subrounded, sandsotne, with subrounded to well-rounded limestone and some subangular to rounded quartz and ironstone Sand: medium, angular to subrounded, quartz with lithic grains	4.2	15.7
Clay associated with Fluvio-glacial and Older River Sand and Gravel	Clay, laminated with silt, sand and gravel seams	2.8	18.5
Upper Marl	Marl, reddish brown	0.5+	19.0

GRADING

Mean for deposit percentages			Depth below surface (m)	percentages						
Fines	Sand	Gravel		Fines	Sand			Gravel		
				$-\frac{1}{16}$	$+\frac{1}{16}-\frac{1}{4}$	$+\frac{1}{4}-1$	+1 -4	+4 -16	+16 -64	+64 mm
6	69	25	11.5-12.7	7	15	35	9	28	6	0
			12.7-13.7	7	13	58	16	6	0	0
			13.7-14.7	5	9	24	17	14	29	2
			14.7-15.7	6	11	40	28	6	9	0
			Mean	6	12	40	17	14	11	0

SE 45 SW 89 **4435 5099** **Ainsty Beck** **Block D**

Surface level +20.7 m
Water struck at + 18.1 m
December 1977

Overburden 3.0 m
Mineral 4.1 m
Waste 9.4 m
Mineral 7.6 m
Bedrock 0.8 m+

LOG

Geological classification	Lithology	Thickness m	Depth m
	Soil	0.4	0.4
Peat	Peat, dark brown, with roots and wood fragments	2.4	2.8
Till	Clay, sandy, pebbly, brown	0.2	3.0
Glacial Sand and Gravel	**a** Pebbly sand Gravel: fine and coarse, subangular to rounded, sandstone with limestone, and some angular chert and rounded ironstone Sand: fine, subangular to rounded, quartz with lithic grains	4.1	7.1
Till	Clay, sandy, pebbly, brown, becoming silty below 11.3m and laminated below 14.5m	7.6	14.7
	Silt, sandy, pebbly, brown	0.7	15.4
Clay associated with Fluvio-glacial and Older River Sand and Gravel	Clay, silty, pebbly, poorly laminated, dark brown	1.1	16.5
Fluvio-glacial and Older River Sand and Gravel	**b** Pebbly sand Gravel: fine and coarse, subangular to rounded sandstone and rounded to well-rounded limestone with some quartz Sand: medium and fine, subangular to rounded quartz with lithic grains	7.6	24.1
Sherwood Sandstone Group	Marl, reddish brown	0.8+	24.9

GRADING

	Mean for deposit percentages			Depth below surface (m)	percentages						
	Fines	Sand	Gravel		Fines	Sand			Gravel		
					$-\frac{1}{16}$	$+\frac{1}{16}-\frac{1}{4}$	$+\frac{1}{4}-1$	+1 -4	+4 -16	+16 -64	+64 mm
a	8	81	11	3.0-5.0	9	46	41	4	0	0	0
				5.0-6.0	6	37	29	8	11	9	0
				6.0-7.1	6	42	15	14	11	12	0
				Mean	8	42	31	8	6	5	0
b	6	72	22	16.5-17.5	11	58	26	2	2	1	0
				17.5-18.5	13	49	34	4	0	0	0
				18.5-19.5	5	37	35	10	10	3	0
				19.5-21.0	3	20	34	14	18	11	0
				21.0-22.3	4	10	35	17	21	13	0
				22.3-23.7	3	10	45	12	15	15	0
				23.7-24.1	1	5	29	13	19	31	2
				Mean	6	27	35	10	13	9	0
a + b	6	76	18	Mean	6	32	34	10	10	8	0

SE 45 SE 22 **4522 5464** **Old Thornville** **Block D**

Surface level +24.2 m
Water struck at +17.7 m
November 1977

Overburden 14.0 m
Mineral 5.3 m
Bedrock 3.0 m+

LOG

Geological classification	Lithology	Thickness m	Depth m
	Soil	1.0	1.0
Till	Clay, sandy, pebbly, brown, sporadic thin seams of fine-grained sand	10.0	11.0
Clay associated with Fluvio-glacial and Older River Sand and Gravel	Clay, laminated with partings of fine-grained sand and silt, sporadic thin seams of fine-grained sand	3.0	14.0
Fluvio-glacial and Older River Sand and Gravel	**a** 'Clayey' sand: medium, subangular to subrounded, equant, quartz with lithic grains	2.2	16.2
	b Sandy gravel Gravel: coarse, rounded to well-rounded, equant, sandstone with subrounded to rounded, tabular and equant, limestone Sand: medium, subangular to subrounded, equant, quartz with lithic grains	3.1	19.3
Sherwood Sandstone Group	Sandstone, brown	3.0+	22.3

GRADING

	Mean for deposit percentages			Depth below surface (m)	percentages						
	Fines	Sand	Gravel		Fines	Sand			Gravel		
					$-\frac{1}{16}$	$+\frac{1}{16}-\frac{1}{4}$	$+\frac{1}{4}-1$	+1 -4	+4 -16	+16 -64	+64 mm
a	12	84	4	14.0-15.0	14	36	40	8	2	0	0
				15.0-16.2	9	21	58	6	3	3	0
				Mean	12	27	50	7	2	2	0
b	5	53	42	16.2-17.1	9	8	21	11	17	34	0
				17.1-18.1	5	8	32	15	14	25	1
				18.1-19.3	2	7	33	20	18	14	6
				Mean	5	8	29	16	16	23	3
a + b	8	65	27	Mean	8	15	38	12	11	14	2

SE 45 SE 23 **4725 5471** **Mill Farm** **Block B**

Surface level +14.3 m
Water struck at +12.2 m
November 1977

Overburden 1.7 m
Mineral 1.0 m
Waste 5.1 m
Mineral 8.2 m
Bedrock 1.3 m+

LOG

Geological classification	Lithology	Thickness m	Depth m
Glacial Lake Deposits	Soil, silty, sandy with pebbles	0.3	0.3
	Clay, silty becoming sandy, mottled, brown	1.4	1.7
	a 'Very clayey' pebbly sand Gravel: fine, subangular to rounded, equant, sandstone with limestone Sand: fine, subangular to subrounded, equant, quartz with lithic grains	1.0	2.7
	Clay, laminated, silty with partings of fine-grained sand and sporadic pebbles	4.5	7.2
Till	Clay, sandy, pebbly, dark grey	0.6	7.8
Fluvio-glacial and Older River Sand and Gravel	**b** 'Clayey' sandy gravel Gravel: coarse, subangular to subrounded, tabular and equant sandstone with, rounded to well-rounded, equant, limestone and some quartz and angular chert Sand: fine, subangular to subrounded, equant, quartz with some lithic grains	8.2	16.0
Sherwood Sandstone Group	Sandstone, reddish brown	1.3+	17.3

GRADING

	Mean for deposit percentages			Depth below surface (m)	percentages						
	Fines	Sand	Gravel		Fines	Sand			Gravel		
					$-\frac{1}{16}$	$+\frac{1}{16}-\frac{1}{4}$	$+\frac{1}{4}-1$	+1 -4	+4 -16	+16 -64	+64 mm
a	37	53	10	1.7-2.7	37	33	13	7	7	3	0
b	15	54	31	7.8-10.8	21	53	10	4	5	7	0
				10.8-12.6	11	43	13	7	7	13	6
				12.6-13.6	27	7	11	11	21	23	0
				13.6-14.4	7	28	25	15	11	14	0
				14.4-16.0	5	14	8	7	10	51	5
				Mean	15	35	12	7	9	20	2
a + b	18	53	29	Mean	18	35	11	7	9	18	2

SE 45 SE 24 **4820 5410** **Wilstrop Grange** **Block B**

Surface level +15.6 m
Water struck at +7.7 m
December 1977

Overburden 7.9 m
Mineral 7.6 m
Bedrock 1.0 m+

LOG

Geological classification	Lithology	Thickness m	Depth m
	Soil	0.3	0.3
Glacial Lake Deposits	Clay, sandy, mottled brown	1.6	1.9
	Clay, poorly laminated to 3.0 m then becoming well laminated, brown, thin seams of fine-grained sand	6.0	7.9
?Sandy Till	**a** 'Very clayey' pebbly sand Gravel: coarse, subangular to subrounded, sandstone with limestone Sand: fine, subangular to subrounded, quartz with lithic grains	1.3	9.2
Fluvio-glacial and Older River Sand and Gravel	**b** Gravel Gravel: coarse and fine, subangular to rounded, equant sandstone and subrounded to rounded tabular limestone Sand: medium, subangular to subrounded, quartz with lithic grains	6.3	15.5
Sherwood Sandstone Group	Sandstone, brown	1.0+	16.5

GRADING

	Mean for deposit percentages			Depth below surface (m)	percentages						
	Fines	Sand	Gravel		Fines	Sand			Gravel		
					$-\frac{1}{16}$	$+\frac{1}{16}-\frac{1}{4}$	$+\frac{1}{4}-1$	+1 -4	+4 -16	+16 -64	+64 mm
a	28	61	11	7.9-9.2	28	40	18	3	4	7	0
b	3	47	50	9.2-10.2	13	29	22	7	15	14	0
				10.2-11.2	4	10	37	12	20	17	0
				11.2-12.2	0	6	23	14	25	29	3
				12.2-13.2	2	4	14	22	33	24	1
				13.2-14.2	0	3	20	16	28	33	0
				14.2-15.5	1	3	22	16	20	36	2
				Mean	3	9	23	15	23	26	1
a + b	7	49	44	Mean	7	14	22	13	20	23	1

SE 45 SE 25 **4897 5468** **Shaw's Rash** **Block B**

Surface level +14.4 m
Water struck at +5.6 m
December 1977

Overburden 8.8 m
Mineral 4.8 m
Waste 0.5 m
Bedrock 1.0 m+

LOG

Geological classification	Lithology	Thickness m	Depth m
	Soil	0.3	0.3
Glacial Lake Deposits	Clay, laminated, mottled to 2.0 m then dark brown	7.3	7.6
Till	Clay, sandy, brown, sporadic pebbles	1.2	8.8
Fluvio-glacial and Older River Sand and Gravel	**a** 'Clayey' sand: fine, subangular to subrounded, equant, quartz with lithic grains	2.2	11.0
	b Gravel Gravel: coarse, subangular to rounded, equant sandstone with some subrounded to rounded, equant quartz Sand: fine and medium, subangular to subrounded, equant, quartz with some lithic grains	2.6	13.6
Clay associated with Fluvio-glacial and Older River Sand and Gravel	Clay, sandy, greyish brown	0.5	14.1
Sherwood Sandstone Group	Sandstone, brown	1.0+	15.1

GRADING

	Mean for deposit percentages			Depth below surface (m)	percentages						
	Fines	Sand	Gravel		Fines	Sand			Gravel		
					$-\frac{1}{16}$	$+\frac{1}{16}-\frac{1}{4}$	$+\frac{1}{4}-1$	+1 -4	+4 -16	+16 -64	+64 mm
a	14	85	1	8.8-10.0	18	68	12	2	0	0	0
				10.0-11.0	8	64	24	2	2	0	0
				Mean	14	66	17	2	1	0	0
b	4	31	65	11.0-12.1	5	14	12	7	12	43	7
				12.1-13.6	3	9	10	10	19	45	4
				Mean	4	11	11	9	16	44	5
a + b	9	55	36	Mean	9	36	14	5	9	24	3

SE 45 SE 26 4540 5331 Crossgate Farm **Block D**

Surface level +18.2 m
Water struck at +8.4 m
December 1977

Overburden 13.1 m
Mineral 5.2 m
Bedrock 3.7 m+

LOG

Geological classification	Lithology	Thickness m	Depth m
	Made Ground	0.4	0.4
Glacial Lake Deposits	Clay, sandy, pebbly, mottled, slightly laminated	2.9	3.3
Till	Clay, sandy, pebbly, brown	9.8	13.1
Fluvio-glacial and Older River Sand and Gravel	**a** Pebbly sand Gravel: fine, rounded to well-rounded, limestone and sandstone with shale Sand: medium, quartz with lithic grains	2.1	15.2
	b Gravel Gravel: coarse, rounded to well-rounded, limestone and sandstone with shale and some quartz and volcanics Sand: medium, quartz with lithic grains	3.1	18.3
Sherwood Sandstone Group	Sandstone, reddish brown, weathered	3.7+	22.0

GRADING

	Mean for deposit percentages			Depth below surface (m)	percentages						
	Fines	Sand	Gravel		Fines	Sand			Gravel		
					$-\frac{1}{16}$	$+\frac{1}{16}-\frac{1}{4}$	$+\frac{1}{4}-1$	+1 -4	+4 -16	+16 -64	+64 mm
a	7	77	16	13.1-14.0	9	17	48	15	11	0	0
				14.0-15.2	6	10	38	26	16	4	0
				Mean	7	13	43	21	14	2	0
b	2	48	50	15.2-16.0	4	11	32	10	12	31	0
				16.0-17.0	1	13	26	11	16	33	0
				17.0-18.3	2	10	18	16	14	40	0
				Mean	2	11	24	13	14	36	0
a + b	4	60	36	Mean	4	12	32	16	14	22	0

SE 45 SE 27 **4599 5396** **Nethercarr** **Block D**

Surface level +15.2 m
Water struck at +8.3 m
November 1977

Overburden 1.6 m
Mineral 3.4 m
Waste 2.0 m
Mineral 5.0 m
Bedrock 0.6 m+

LOG

Geological classification	Lithology	Thickness m	Depth m
	Soil	0.3	0.3
River Terrace Deposits, undifferentiated	Clay, sandy, silty, brown	1.3	1.6
	a 'Very clayey' sandy gravel Gravel: coarse and fine, subrounded to rounded, equant sandstone with limestone Sand: fine, subangular to subrounded, equant quartz with some lithic grains Fines: thin seams of sandy clay at 2.9 m and 4.2 m	3.4	5.0
Till	Clay, sandy, pebbly, brown	2.0	7.0
Fluvio-glacial and Older River Sand and Gravel	**b** Pebbly sand Gravel: fine, subangular to rounded, equant, sandstone with angular chert and rounded limestone and some ironstone, shale and quartz Sand: medium, subangular to subrounded, equant quartz with some lithic grains	2.0	9.0
	c Sandy gravel Gravel: coarse and fine, subrounded to rounded, equant sandstone with some limestone Sand: medium, subangular to rounded, equant quartz with some lithic grains	3.0	12.0
Sherwood Sandstone Group	Sandstone	0.6+	12.6

GRADING

	Mean for deposit percentages			Depth below surface (m)	percentages						
	Fines	Sand	Gravel		Fines	Sand			Gravel		
					$-\frac{1}{16}$	$+\frac{1}{16}-\frac{1}{4}$	$+\frac{1}{4}-1$	+1 -4	+4 -16	+16 -64	+64 mm
a	23	53	24	1.6-2.6	10	8	22	9	21	30	0
				2.6-3.6	32	34	9	4	8	6	7
				3.6-5.0	26	53	13	2	4	2	0
				Mean	23	34	14	5	10	12	2
b	6	89	5	7.0-8.0	6	20	55	17	2	0	0
				8.0-9.0	6	17	53	16	8	0	0
				Mean	6	19	54	16	5	0	0
c	1	53	46	9.0-10.1	1	7	32	12	17	31	0
				10.1-11.1	1	4	34	17	17	19	8
				11.1-12.0	0	7	30	17	19	27	0
				Mean	1	6	32	15	17	26	3
a+b+c	11	61	28	Mean	11	20	30	11	12	14	2

SE 45 SE 28 **4747 5356** **Tockwith Moor** **Block B**

Surface level +15.5 m
Water struck at +6.2 m
September 1977

Overburden 9.3 m
Mineral 8.2 m
Bedrock 1.1 m+

LOG

Geological classification	Lithology	Thickness m	Depth m
	Soil	0.3	0.3
Glacial Lake Deposits	Clay, laminated, brown, with grey silt partings	6.9	7.2
	'Very clayey' sand: fine, quartz with lithic grains	1.4	8.6
Till	Clay, pebbly, brown	0.7	9.3
Fluvio-glacial and Older River Sand and Gravel	**a** 'Clayey' pebbly sand Gravel: fine, sandstone and limestone Sand: medium, quartz with lithic grains Fines: thin seams of sandy clay at 10.7 m	5.9	15.2
	b Gravel Gravel: fine, sandstone and limestone with shale and quartz and some chert Sand: medium, quartz with lithic grains	2.3	17.5
Sherwood Sandstone Group	Sandstone, reddish brown, weathered	1.1+	18.6

GRADING

	Mean for deposit percentages			Depth below surface (m)	percentages						
	Fines	Sand	Gravel		Fines	Sand			Gravel		
					$-\frac{1}{16}$	$+\frac{1}{16}-\frac{1}{4}$	$+\frac{1}{4}-1$	+1 -4	+4 -16	+16 -64	+64 mm
a	14	81	5	9.3-10.3	28	43	24	5	0	0	0
				10.3-12.0	13	37	32	8	10	0	0
				12.0-13.1	11	37	50	2	0	0	0
				13.1-14.1	11	37	50	2	0	0	0
				14.1-15.2	6	21	48	12	10	3	0
				Mean	14	35	40	6	4	1	0
b	4	47	49	15.2-16.5	4	13	27	14	24	18	0
				16.5-17.5	5	8	13	17	34	23	0
				Mean	4	11	21	15	29	20	0
a + b	11	72	17	Mean	11	28	35	9	11	6	0

SE 45 SE 29 **4931 5359** **White Sike** **Block B**

Surface level +13.7 m
Water struck at +5.7 m
September 1977

Overburden 9.6 m
Mineral 10.9 m
Bedrock 0.4 m+

LOG

Geological classification	Lithology	Thickness m	Depth m
	Soil	1.2	1.2
Glacial Lake Deposits	Clay, silty, laminated, brown, with partings of greyish brown silt and fine-grained sand	6.9	8.1
?Till	Clay, sandy, pebbly, brown	1.5	9.6
Fluvio-glacial and Older River Sand and Gravel	**a** 'Clayey' sand: fine, quartz with lithic grains	5.4	15.0
	b Sandy gravel Gravel: coarse, rounded, sandstone and limestone with some vein quartz and shale Sand: medium, quartz with lithic grains	5.5	20.5
Sherwood Sandstone Group	Sandstone, reddish brown	0.4+	20.9

GRADING

	Mean for deposit percentages			Depth below surface (m)	percentages						
	Fines	Sand	Gravel		Fines	Sand			Gravel		
					$-\frac{1}{16}$	$+\frac{1}{16}-\frac{1}{4}$	$+\frac{1}{4}-1$	+1 -4	+4 -16	+16 -64	+64 mm
a	19	81	0	9.6-10.6	18	73	9	0	0	0	0
				10.6-11.6	25	57	17	1	0	0	0
				11.6-12.6	26	48	23	3	0	0	0
				12.6-13.6	18	49	31	2	0	0	0
				13.6-15.0	13	40	44	3	0	0	0
				Mean	19	53	26	2	0	0	0
b	4	66	30	15.0-16.0	4	13	32	16	19	16	0
				16.0-17.0	2	16	44	7	14	17	0
				17.0-18.5	3	13	37	11	15	17	4
				18.5-19.7	3	17	24	19	13	24	0
				19.7-20.5	12	50	34	4	0	0	0
				Mean	4	20	34	12	13	16	1
a + b	12	73	15	Mean	12	36	30	7	6	8	1

SE 45 SE 30 **4581 5220** **Tockwith** **Block D**

Surface level +21.0 m
Water struck at +10.0 m
December 1977

Waste 15.7 m
Bedrock 1.3 m+

LOG

Geological classification	Lithology	Thickness m	Depth m
Made ground	Soil	0.3	0.3
Till	Clay, sandy, pebbly, mottled, brown below 6.0 m	10.7	11.0
Glacial Sand and Gravel	'Clayey' sand: medium, quartz with lithic grains	1.9	12.9
Till	Clay, pebbly	1.1	14.0
Fluvio-glacial and Older River Sand and Gravel	'Very clayey' sand: fine, quartz with lithic grains	1.7	15.7
Sherwood Sandstone Group	Mudstone and sandstone, weathered	1.3+	17.0

SE 45 SE 31 **4667 5291** **Tockwith** **Block D**

Surface level +17.0 m
Water struck at +8.5 m
September 1977

Overburden 1.7 m
Mineral 4.0 m
Waste 2.8 m
Mineral 7.3 m
Bedrock 1.1 m+

LOG

Geological classification	Lithology	Thickness m	Depth m
	Soil	0.2	0.2
Glacial Lake Deposits	Clay, silty, poorly laminated, mottled	1.5	1.7
Sandy Till	**a** 'Very clayey' pebbly sand Gravel: fine, sandstone with limestone and some shale and chert Sand: fine, quartz with lithic grains	4.0	5.7
Clay associated with Fluvio-glacial and Older River Sand and Gravel	Clay, laminated, greyish brown, with fine-grained sand partings	2.8	8.5
Fluvio-glacial and Older River Sand and Gravel	**b** 'Very clayey' pebbly sand Gravel: fine and coarse, sandstone and limestone with shale Sand: fine, quartz with lithic grains	7.3	15.8
Sherwood Sandstone Group	Sandstone, reddish brown	1.1+	16.9

GRADING

	Mean for deposit percentages			Depth below surface (m)	percentages						
	Fines	Sand	Gravel		Fines	Sand			Gravel		
					$-\frac{1}{16}$	$+\frac{1}{16}-\frac{1}{4}$	$+\frac{1}{4}-1$	+1 -4	+4 -16	+16 -64	+64 mm
a	30	62	8	1.7-2.7	27	56	12	2	3	0	0
				2.7-3.7	29	54	9	3	3	2	0
				3.7-4.7	27	43	12	3	9	6	0
				4.7-5.7	38	41	10	3	7	1	0
				Mean	30	48	11	3	6	2	0
b	30	64	6	8.5-11.5	35	27	13	11	4	10	0
				11.5-13.7	32	55	9	4	0	0	0
				13.7-14.7	23	55	19	3	0	0	0
				14.7-15.8	19	59	19	3	0	0	0
				Mean	30	44	13	7	2	4	0
a + b	30	63	7	Mean	30	46	12	5	3	4	0

SE 45 SE 32 **4837 5298** **White Sike Close** **Block B**

Surface level +15.9 m
Water struck at +6.6 m
September 1977

Overburden 6.5 m
Mineral 4.8 m
Bedrock 1.5 m+

LOG

Geological classification	Lithology	Thickness m	Depth m
	Soil	0.8	0.8
Glacial Lake Deposits	Clay, poorly laminated in part, mottled	5.7	6.5
Fluvio-glacial and Older River Sand and Gravel (Upper part may be ?Sandy Till)	'Very clayey' pebbly sand Gravel: fine, sandstone and limestone with some chert, shale and vein quartz Sand: fine, quartz with lithic grains	4.8	11.3
Sherwood Sandstone Group	Sandstone, brown	1.5+	12.8

GRADING

Mean for deposit percentages			Depth below surface (m)	percentages						
Fines	Sand	Gravel		Fines	Sand			Gravel		
				$-\frac{1}{16}$	$+\frac{1}{16}-\frac{1}{4}$	$+\frac{1}{4}-1$	+1 -4	+4 -16	+16 -64	+64 mm
27	67	6	6.5-7.6	39	41	10	3	7	0	0
			7.6-9.3	21	56	15	3	4	1	0
			9.3-10.3	26	48	16	4	6	0	0
			10.3-11.3	26	48	15	6	4	1	0
			Mean	27	49	14	4	5	1	0

SE 45 SE 33 **4919 5245** **Moor Lane**

Surface level +14.6 m
Water struck at +7.2 m
September 1977

Waste 11.9 m
Bedrock 1.1 m+

LOG

Geological classification	Lithology	Thickness m	Depth m
	Soil	0.7	0.7
Glacial Lake Deposits	Clay, silty, laminated brown, with grey silt partings	3.3	4.0
	'Very clayey' pebbly sand Gravel: fine, sandstone and limestone Sand: fine, quartz with lithic grains	1.0	5.0
Till	Silt, sandy, brown	1.0	6.0
	Clay, sandy, pebbly, brown	3.9	9.9
Sandy Till	'Very clayey' pebbly sand Gravel: coarse, sandstone and limestone Sand: fine, quartz with lithic grains	2.0	11.9
Sherwood Sandstone Group	Sandstone, reddish brown, weathered	1.1+	13.0

SE 45 SE 34 **4553 5130** **South Field**

Surface level +25.9 m
Water struck at +21.4 m
December 1977

Waste 4.5 m
Bedrock 3.6 m+

LOG

Geological classification	Lithology	Thickness m	Depth m
	Made ground	0.3	0.3
Till	Clay, sandy, pebbly, brown; sporadic lenses of sand below 2.6 m	4.2	4.5
Sherwood Sandstone Group	Sandstone and mudstone, reddish brown, upper surface weathered	3.6+	8.1

SE 45 SE 35 **4644 5183** **Mill Field** **Block D**

Surface level +22.6 m
Water struck at +11.9 m
November 1977

Overburden 1.6 m
Mineral 8.0 m
Waste 1.0 m
Mineral 1.3 m
Waste 2.6 m
Bedrock 0.9 m+

LOG

Geological classification	Lithology	Thickness m	Depth m
	Soil	0.5	0.5
Till	Clay, sandy, pebbly, brown	1.1	1.6
Sandy Till	**a** 'Very clayey' pebbly sand Gravel: fine and coarse, subrounded to rounded, equant sandstone with limestone Sand: fine, subangular to rounded, equant quartz, with lithic grains	8.0	9.6
Clay associated with Fluvio-glacial and Older River Sand and Gravel	Silt, sandy, grey, laminated at base	1.0	10.6
Fluvio-glacial and Older River Sand and Gravel	**b** 'Very clayey' sand: fine, subangular to rounded, equant quartz with some lithic grains	1.3	11.9
?Till	Clay, sandy, pebbly, brown	2.6	14.5
Sherwood Sandstone Group	Sandstone	0.9+	15.4

GRADING

	Mean for deposit percentages			Depth below surface (m)	percentages						
	Fines	Sand	Gravel		Fines	Sand			Gravel		
					$-\frac{1}{16}$	$+\frac{1}{16}-\frac{1}{4}$	$+\frac{1}{4}-1$	+1 -4	+4 -16	+16 -64	+64 mm
a	33	58	9	1.6-2.6	29	43	13	4	6	5	0
				2.6-3.6	29	45	14	4	4	4	0
				3.6-4.6	35	44	11	3	4	3	0
				4.6-5.6	30	45	11	3	3	5	3
				5.6-6.6	33	43	12	3	4	5	0
				6.6-7.6	34	42	12	3	4	5	0
				7.6-8.6	39	43	12	3	3	0	0
				8.6-9.6	39	37	9	2	3	6	4
				Mean	33	43	12	3	4	4	1
b	23	77	0	10.6-11.9	23	71	6	0	0	0	0
a + b	32	61	7	Mean	32	47	11	3	3	3	1

SE 45 SE 36 **4755 5143** **Bilton Grange**

Surface level +26.9 m
Water struck at +14.3 m
November 1977

Waste 12.6 m
Bedrock 2.3 m+

LOG

Geological classification	Lithology	Thickness m	Depth m
	Soil	0.3	0.3
Till	Clay, sandy, pebbly, brown; sporadic thin seams of pebbly sand, poorly laminated at 4.8 m	12.3	12.6
Sherwood Sandstone Group	Sandstone and marl, brown, weathered	2.3+	14.9

SE 45 SE 37 **4830 5106** **Church Field**

Surface level +37.7 m
Water not encountered
September 1977

Waste 14.1 m
Bedrock 2.0 m+

LOG

Geological classification	Lithology	Thickness m	Depth m
	Soil	0.3	0.3
Till	Clay, sandy, silty, pebbly, brown, poorly laminated at 9.6 m; sporadic sand lenses below 10.0 m	13.8	14.1
Sherwood Sandstone Group	Sandstone and marl, brown, weathered	2.0+	16.1

SE 45 SE 38 **4645 5086** **West Grange**

Surface level +32.6 m
Water struck at +26.3 m
September 1977

Waste 4.4 m
Bedrock 1.9 m+

LOG

Geological classification	Lithology	Thickness m	Depth m
	Soil	0.5	0.5
Till	Clay, sandy, pebbly, brown, poorly laminated at 2.4 m	3.9	4.4
Sherwood Sandstone Group	Sandstone and marl, brown,weathered	1.9+	6.3

REFERENCES

ALLEN, V. T. 1936. Terminology of medium-grained sediments. **Rep. Natl. Res. Counc., Washington, 1935–1936, App. 1, Rep. Comm. Sediment.**, 18-47.

ARCHER, A. A. 1969. Background and problems of an assessment of sand and gravel resources in the United Kingdom. **Proc. 9th Commonw. Min. & Metall. Congr., 1969,** Vol. 2: Mining and petroleum geology, 495-508.

— 1970a. Standardisation of the size classification of naturally occurring particles. **Geotechnique**, Vol. 20, 103-107.

— 1970b. Making the most of metrication. **Quarry Managers' J.**, Vol. 54, No. 6, 223-227.

ATTERBERG, A. 1905. Die rationelle Klassifikation der Sande und Kiese. **Chem. Z.**, Vol. 29, 195-198.

BRITISH STANDARDS INSTITUTION. 1967. **B.S.1377: Methods of testing soils for civil engineering purposes.** (London: British Standards Institution.)

BUREAU OF MINES AND GEOLOGICAL SURVEY. 1948. **Pp. 14-17** in **Mineral resources of the United States.** (Washington, DC: Public Affairs Press.)

EDWARDS, W. 1936. A Pleistocene strand line in the Vale of York. **Proc. Yorkshire Geol. Soc.**, Vol. 23, 103-118.

— MITCHELL, G. H., and WHITEHEAD, T. H. 1950. Geology of the district north and east of Leeds. **Mem. Geol. Surv. G. B.**, Sheet 70, 93.

GAUNT, G. D. 1974. A radiocarbon date relating to Lake Humber. **Proc. Yorkshire Geol. Soc.**, Vol. 40, 195-197.

— 1976. The Devensian maximum ice limit in the Vale of York. **Proc. Yorkshire Geol. Soc.**, Vol. 40, 631-637.

— JARVIS, R. A. and MATTHEWS, B. 1971. The late Weichselian sequence in the Vale of York. **Proc. Yorkshire Geol. Soc.**, Vol. 38, 281-284.

HARRIS, P. M., THURRELL, R. G., HEALING, R. A., and ARCHER, A. A. 1974. Aggregates in Britain. **Proc. R. Soc.**, Ser. A, Vol. 339, 329-353.

HULL, J. H. 1981. Methods of calculating the volume of resources of sand and gravel. **Appendix** (pp. 192-193) to **THURRELL, R. G.** 1981. Quarry resources and reserves: the identification of bulk mineral resources: the contribution of the Institute of Geological Sciences. **Quarry Management**, for March 1981, 181-193.

LANE, E. W., and others. 1947. Report of the sub-committee on sediment terminology. **Trans. Am. Geophys. Union**, Vol. 28, 936-938.

PETTIJOHN, F. J. 1957. **Sedimentary rocks.** 2nd edition. (London: Harper and Row.)

THURRELL, R. G. 1971. The assessment of mineral resources with particular reference to sand and gravel. **Quarry Managers' J.**, Vol. 55, 19-25.

— 1981. Quarry resources and reserves: the identification of bulk mineral resources: the contribution of the Institute of Geological Sciences. **Quarry Management**, for March 1981, 181-193.

TWENHOFEL, W. H. 1937. Terminology of the fine-grained mechanical sediments. **Rep. Natl. Res. Counc., Washington, 1936–37,** App. 1, Rep. Comm. Sediment., 81-104.

UDDEN, J. A. 1914. Mechanical composition of clastic sediments. **Bull. Geol. Soc. Am.,** Vol. 25, 655-744.

WENTWORTH, C. K. 1922. A scale of grade and class terms for clastic sediments. **J. Geol.**, Vol. 30, 377-392.

— 1935. The terminology of coarse sediments. **Bull. Natl. Res. Counc. Washington**, No. 98, 225-246.

WILLMAN, H. B. 1942. Geology and mineral resources of the Marseilles, Ottawa and Streator quadrangles. **Bull. Illinois State Geol. Surv.**, No. 66, 343-344.

The following reports of the Institute relate particularly to bulk mineral resources

Reports of the Institute of Geological Sciences

Assessment of British Sand and Gravel Resources

1 The sand and gravel resources of the country south-east of Norwich, Norfolk: Resource sheet TG 20.
E. F. P. Nickless.
Report 71/20 ISBN 0 11 880216 X £1.15

2 The sand and gravel resources of the country around Witham, Essex: Resource sheet TL 81. H. J. E. Haggard.
Report 72/6 ISBN 0 11 880588 6 £1.20

3 The sand and gravel resources of the area south and west of Woodbridge, Suffolk: Resource sheet TM 24.
R. Allender and S. E. Hollyer.
Report 72/9 ISBN 0 11 880596 7 £1.70

4 The sand and gravel resources of the country around Maldon, Essex: Resource sheet TL 80. J. D. Ambrose.
Report 73/1 ISBN 0 11 880600 9 £1.20

5 The sand and gravel resources of the country around Hethersett, Norfolk: Resource sheet TG 10.
E. F. P. Nickless.
Report 73/4 ISBN 0 11 880606 8 £1.60

6 The sand and gravel resources of the country around Terling, Essex: Resource sheet TL 71. C. H. Eaton.
Report 73/5 ISBN 0 11 880608 4 £1.20

7 The sand and gravel resources of the country around Layer Breton and Tolleshunt D'Arcy, Essex: Resource sheet TL 91 and part of TL 90. J. D. Ambrose.
Report 73/8 ISBN 0 11 880614 9 £1.30

8 The sand and gravel resources of the country around Shotley and Felixstowe, Suffolk: Resource sheet TM 23.
R. Allender and S. E. Hollyer.
Report 73/13 ISBN 0 11 880625 4 £1.60

9 The sand and gravel resources of the country around Attlebridge, Norfolk: Resource sheet TG 11.
E. F. P. Nickless.
Report 73/15 ISBN 0 11 880658 0 £1.85

10 The sand and gravel resources of the country west of Colchester, Essex: Resource sheet TL 92. J. D. Ambrose.
Report 74/6 ISBN 0 11 880671 8 £1.45

11 The sand and gravel resources of the country around Tattingstone, Suffolk: Resource sheet TM 13. S. E. Hollyer.
Report 74/9 ISBN 0 11 880675 0 £1.95

12 The sand and gravel resources of the country around Gerrards Cross, Buckinghamshire: Resource sheet SU 99, TQ 08 and TQ 09. H. C. Squirrell.
Report 74/14 ISBN 0 11 880710 2 £2.20

Mineral Assessment Reports

13 The sand and gravel resources of the country east of Chelmsford, Essex: Resource sheet TL 70. M. R. Clarke.
ISBN 0 11 880744 7 £3.50

14 The sand and gravel resources of the country east of Colchester, Essex: Resource sheet TM 02. J. D. Ambrose.
ISBN 0 11 880745 5 £3.25

15 The sand and gravel resources of the country around Newton on Trent, Lincolnshire: Resource sheet SK 87.
D. Price.
ISBN 0 11 880746 3 £3.00

16 The sand and gravel resources of the country around Braintree, Essex: Resource sheet TL 72. M. R. Clarke.
ISBN 0 11 880747 1 £3.50

17 The sand and gravel resources of the country around Besthorpe, Nottinghamshire: Resource sheet SK 86 and part of SK 76. J. R. Gozzard.
ISBN 0 11 880748 X £3.00

18 The sand and gravel resources of the Thames Valley, the country around Cricklade, Wiltshire: Resource sheet SU 09/19 and parts of SP 00/10. P. R. Robson.
ISBN 0 11 880749 8 £3.00

19 The sand and gravel resources of the country south of Gainsborough, Lincolnshire: Resource sheet SK 88 and part of SK 78. J. H. Lovell.
ISBN 0 11 880750 1 £2.50

20 The sand and gravel resources of the country east of Newark upon Trent, Nottinghamshire: Resource sheet SK 85.
J. R. Gozzard.
ISBN 0 11 880751 X £2.75

21 The sand and gravel resources of the Thames and Kennet Valleys, the country around Pangbourne, Berkshire: Resource sheet SU 67. H. C. Squirrell.
ISBN 0 11 880752 8 £3.25

22 The sand and gravel resources of the country north-west of Scunthorpe, Humberside: Resource sheet SE 81.
J. W. C. James.
ISBN 0 11 880753 6 £3.00

23 The sand and gravel resources of the Thames Valley, the country between Lechlade and Standlake: Resource sheet SP 30 and parts of SP 20, SU 29 and SU 39. P. Robson.
ISBN 0 11 881252 1 £7.25

24 The sand and gravel resources of the country around Aldermaston, Berkshire: Resource sheet SU 56 and SU 66. H. C. Squirrell.
ISBN 0 11 881253 X £5.00

25 The celestite resources of the area north-east of Bristol: Resource sheet ST 68 and parts of ST 59, 69, 79, 58, 78, 68 and 77. E. F. P. Nickless, S. J. Booth and P. N. Mosley.
ISBN 0 11 881262 9 £5.00

26 The limestone and dolomite resources of the country around Monyash, Derbyshire: Resource sheet SK 16.
F. C. Cox and D. McC. Bridge.
ISBN 0 11 881263 7 £7.00

27 The sand and gravel resources of the country west and south of Lincoln, Lincolnshire: Resource sheets SK 95, SK 96 and SK 97. I. Jackson.
ISBN 0 11 884003 7 £6.00

28 The sand and gravel resources of the country around Eynsham, Oxfordshire: Resource sheet SP 40 and part of SP 41. W. J. R. Harries.
ISBN 0 11 884012 6 £3.00

29 The sand and gravel resources of the country south-west of Scunthorpe, Humberside: Resource sheet SE 80.
J. H. Lovell.
ISBN 0 11 884013 4 £3.50

30 Procedure for the assessment of limestone resources.
F. C. Cox, D. McC. Bridge and J. H. Hull.
ISBN 0 11 884030 4 £1.25

31 The sand and gravel resources of the country west of Newark upon Trent, Nottinghamshire: Resource sheet SK 75.
D. Price and P. J. Rogers.
ISBN 0 11 884031 2 £3.50

32 The sand and gravel resources of the country around Sonning and Henley: Resource sheet SU 77 and SU 78.
H. C. Squirrell.
ISBN 0 11 884032 0 £5.25

33 The sand and gravel resources of the country north of Gainsborough: Resource sheet SK 89. J. R. Gozzard and D. Price
ISBN 0 11 884033 9 £4.50

34 The sand and gravel resources of the Dengie Peninsula, Essex: Resource sheet TL 90, etc. M. B. Simmons.
ISBN 0 11 884081 9 £5.00

35 The sand and gravel resources of the country around Darvel: Resource sheet NS 53, 63, etc. E. F. P. Nickless, A. M. Aitken and A. A. McMillan.
ISBN 0 11 884082 7 £7.00

36 The sand and gravel resources of the country around Southend-on-Sea, Essex: Resource sheets TQ 78, 79 etc. S. E. Hollyer and M. B. Simmons.
ISBN 0 11 884083 5 £7.50

37 The sand and gravel resources of the country around Bawtry, South Yorkshire: Resource sheet SK 69. A. R. Clayton
ISBN 0 11 884053 3 £5.75

38 The sand and gravel resources of the country around Abingdon, Oxfordshire: Resource sheet SU 49, 59, SP 40, 50. C. E. Corser.
ISBN 0 11 884084 5 £5.50

39 The sand and gravel resources of the Blackwater Valley (Aldershot) area: Resource sheet SU 85, 86, parts SU 84, 94, 95, 96. M. R. Clarke, A. J. Dixon and M. Kubala.
ISBN 0 11 884085 1 £7.00

40 The sand and gravel resources of the country west of Darlington, County Durham: Resource sheet NZ 11, 21. A. Smith.
ISBN 0 11 884086 X £5.00

41 The sand and gravel resources of the country around Garmouth, Grampian Region: Resource sheet NJ 36. A. M. Aitken, J. W. Merritt and A. J. Shaw.
ISBN 0 11 884090 8 £8.75

42 The sand and gravel resources of the country around Maidenhead and Marlow: Resource sheet SU 88, parts SU 87, 97, 98. P. N. Dunkley.
ISBN 0 11 884091 6 £5.00

43 The sand and gravel resources of the country around Misterton, Nottinghamshire: Resource sheet SK 79. D. Thomas and D. Price.
ISBN 0 11 884092 4 £5.25

44 The sand and gravel resources of the country around Sedgefield, Durham: Resource sheet NZ 32. M. D. A. Samuel.
ISBN 0 11 884093 2 £5.75

45 The sand and gravel resources of the country around Brampton, Cumbria: Resource sheet NY 55, part 56. I. Jackson.
ISBN 0 11 884094 0 £6.75

46 The sand and gravel resources of the country around Harlow, Essex: Resource sheet TL 41. P. M. Hopson.
ISBN 0 11 884107 6 £9.50

47 The limestone and dolomite resources of the country around Wirksworth, Derbyshire: Resource sheet SK 25, part 35. F. C. Cox and D. J. Harrison.
ISBN 0 11 884108 4 £15.00

48 The sand and gravel resources of the Loddon Valley area: Resource sheet SU 75, 76, parts 64, 65, 66 and 74. M. R. Clarke, E. J. Raynor and R. A. Sobey.
ISBN 0 11 884109 2 £8.75

49 The sand and gravel resources of the country around Lanark, Strathclyde Region: Resource sheet NS 94, part 84. J. L. Laxton and E. F. P. Nickless.
ISBN 0 11 884112 2 £11.00

50 The sand and gravel resources of the country around Fordingbridge, Hampshire: Resource sheet SU 11 and parts of SU 00, 01, 10, 20 and 21. M. Kubala.
ISBN 0 11 884111 4 £7.75

51 The sand and gravel resources of the country north of Bournemouth, Dorset: Resource sheet SU 00, 10, 20, SZ 09, 19 and 29. M. R. Clarke.
ISBN 0 11 884110 6 £9.75

52 The sand and gravel resources of the country between Hatfield Heath and Great Waltham, Essex: Resource sheet TL 51 and 61. R. J. Marks.
ISBN 0 11 884113 0 £8.00

53 The sand and gravel resources of the country around Cottenham, Cambridgeshire: Resource sheet TL 46 and 47. A. J. Dixon.
ISBN 0 11 884114 9 £9.25

54 The sand and gravel resources of the country around Huntingdon and St Ives. Cambridgeshire: Resource sheets TL 16, 17, 26, 27, 36 and 37. R. W. Gatliff.
ISBN 0 11 884115 7 £8.75

55 The sand and gravel resources of the country around Ipswich, Suffolk: Resource sheet TM 14. R. Allender and S. E. Hollyer.
ISBN 0 11 884116 5 £10.00

56 Procedure for the assessment of the conglomerate resources of the Sherwood Sandstone Group. D. P. Piper and P. J. Rodgers.
ISBN 0 11 884143 2 £1.25

57 The conglomerate resources of the Sherwood Sandstone Group of the country around Cheadle, Staffordshire: Resource sheet SK 04. P. J. Rogers, D. P. Piper and T. J. Charsley.
ISBN 0 11 884144 0 *not yet priced*

58 The sand and gravel resources of the country west of Peterhead, Grampian Region: Resource sheet NK 04 and parts of NJ 94 and 95, NK 05, 14 and 15. A. A. McMillan and A. M. Aitken.
ISBN 0 11 884145 9 £12.00

59 The sand and gravel resources of the country around Newbury, Berkshire: Resource sheet SU 46 and 57, parts of SU 36, 37 and 47. J. R. Gozzard.
ISBN 0 11 884146 7 *not yet priced*

60 The sand and gravel resources of the country south-west of Peterborough, in Cambridgeshire and east Northamptonshire: Resource sheet TL 09 and 19 and SP 98 and TL 08. A. M. Harrisson.
ISBN 0 11 884147 5 £15.50

61 The sand and gravel resources of the country north of Wrexham, Clwyd: Resource sheet SJ 35 and part of SJ 25. P. N. Dunkley.
ISBN 0 11 884148 3 £11.75

62 The sand and gravel resources of the country around Dolphinton, Strathclyde Region, and West Linton, Borders Region: Resource sheet NT 04 and 14, parts of NT 05 and 15. A. A. McMillan, J. L. Laxton and A. J. Shaw.
ISBN 0 11 884149 1 £8.00

63 The sand and gravel resources of the valley of the Douglas Water, Strathclyde: Resource sheet NS 83 and parts of NS 82, 92 and 93. A. J. Shaw and E. F. P. Nickless.
ISBN 0 11 884150 5 £11.50

64 The sand and gravel resources of the country between Wallingford and Goring, Oxfordshire: Resource sheet SU 68 and part of SU 58. C. E. Corser.
ISBN 0 11 884151 3 *not yet priced*

65 The sand and gravel resources of the country around Hexham, Northumberland: Resource sheet NY 86 and 96. J. H. Lovell.
ISBN 0 11 884152 1 £7.50

66 The sand and gravel resources of the country west of Chelmsford, Essex: Resource sheet TL 60. P. M. Hopson.
ISBN 0 11 884153 X £8.50

67 The sand and gravel resources of the country around Hatfield and Cheshunt, Hertfordshire: Resource sheet TL 20 and 30, and parts of TQ 29 and 39. J. R. Gozzard.
ISBN 0 11 884167 X £10.00

68 The sand and gravel resources of the country north-east of Halstead, Essex: Resource sheet TL 83. R. J. Marks and J. W. Merritt.
ISBN 0 11 884168 8 £13.25

69 The sand and gravel resources of the country around Welwyn Garden City. Hertfordshire: Resource sheet TL 11 and 21. J. R. Gozzard.
ISBN 0 11 884169 6 £10.50

70 The sand and gravel resources of the country east of Harrogate, North Yorkshire: Resource sheet SE 35. D. L. Dundas.
ISBN 0 11 884170 7 £9.50

71 The sand and gravel resources of the country around Hemel Hempstead, St Albans and Watford: Resource sheet TL 00 and 10, and parts of TQ 09 and 19.
W. J. R. Harries, S. E. Hollyer and P. M. Hopson.
ISBN 0 11 884171 8 *not yet priced*

72 The sand and gravel resources of the country around Bury St Edmunds Suffolk: Resource sheet TL 86.
M. P. Hawkins.
ISBN 0 11 884172 6 £10.50

73 The sand and gravel resources of the country between Ely and Cambridge, Cambridgeshire: Resource sheet TL 56, 57.
A. R. Clayton.
ISBN 0 11 884173 4 £9.50

74 The sand and gravel resources of the country around Blaydon, Tyne and Wear: Resource sheet NZ 06, 16.
J. R. A. Giles.
ISBN 0 11 884174 2 *not yet priced*

75 The sand and gravel resources of the country around Stokesley, North Yorkshire: Resource sheet NZ 40, 50 and parts 41, 51. R. G. Crofts.
ISBN 0 11 884175 0 *not yet priced*

76 The sand and gravel resources of the country around Ellon, Grampian Region: Resource sheets NJ 93 with parts of NJ 82, 83 and 92 and NK 03 with parts of NK 02 and 13.
J. W. Merritt.
ISBN 0 11 884176 9 £15.00

77 The limestone and dolomite resources of the country around Buxton, Derbyshire: Resource sheet SK 07 and parts of SK 06 and 08. D. J. Harrison.
ISBN 0 11 884177 7 *not yet priced*

78 The sand and gravel resources of the country west of Boroughbridge, North Yorkshire: Resource sheet SE 30.
D. A. Abraham.
ISBN 0 11 884178 5 *not yet priced*

79 The limestone and dolomite resources of the country around Bakewell, Derbyshire: Resource sheet SK 26 and part of SK 27. D. McC. Bridge and J. R. Gozzard.
ISBN 0 11 884179 3 *not yet priced*

80 The sand and gravel resources of the country between Stamford and Peterborough: Resource sheet TF 00 and 10.
S. J. Booth.
ISBN 0 11 884180 7 *not yet priced*

81 The sand and gravel resources of the Thames and Thame valleys, the country around Dorchester and Watlington, Oxfordshire: Resource sheet SU 69 and part 59.
C. E. Corser.
ISBN 0 11 884204 8 *not yet priced*

82 The sand and gravel resources of the country around Sible Hedingham, Essex: Resource sheet TL 73. R. J. Marks and D. W. Murray.
ISBN 0 11 884205 6 £10.75

83 The sand and gravel resources of the country around Hollesley, Suffolk: Resource sheet TM 34. S. E. Hollyer and R. Allender.
ISBN 0 11 884206 4 *not yet priced*

84 The sand and gravel resources of the country around Kirk Hammerton, North Yorkshire: Resource sheet SE 45.
J. R. A. Giles.
ISBN 0 11 884207 2 £10.00

Reports of the Institute of Geological Sciences

Other Reports

69/9 Sand and gravel resources of the Inner Moray Firth.
A. L. Harris and J. D. Peacock.
ISBN 0 11 880106 6 35p

70/4 Sands and gravels of the southern counties of Scotland.
G. A. Goodlet.
ISBN 0 11 880105 8 90p

72/8 The use and resources of moulding sand in Northern Ireland. R. A. Old.
ISBN 0 11 881594 0 30p

73/9 The superficial deposits of the Firth of Clyde and its sea lochs. C. E. Deegan, R. Kirby, I. Rae and R. Floyd.
ISBN 0 11 880617 3 95p

77/1 Sources of aggregate in Northern Ireland (2nd edition).
I. B. Cameron.
ISBN 0 11 881279 3 70p

77/2 Sand and gravel resources of the Grampian Region.
J. D. Peacock and others.
ISBN 0 11 881282 3 80p

77/5 Sand and gravel resources of the Fife Region.
M. A. E. Browne.
ISBN 0 11 884004 5 60p

77/6 Sand and gravel resources of the Tayside Region.
I. B. Paterson.
ISBN 0 11 884008 8 £1.40

77/8 Sand and gravel resources of the Strathclyde Region.
I. B. Cameron and others.
ISBN 0 11 884028 2 £2.50

77/9 Sand and gravel resources of the Central Region, Scotland. M. A. E. Browne.
ISBN 0 11 884016 9 £1.35

77/19 Sand and gravel resources of the Borders Region, Scotland. A. D. McAdam.
ISBN 0 11 884025 8 £1.00

77/22 Sand and gravel resources of the Dumfries and Galloway Region of Scotland. I. B. Cameron.
ISBN 0 11 884021 5 £1.20

78/1 Sand and gravels of the Lothian Region of Scotland.
A. D. McAdam.
ISBN 0 11 884042 8 £1.00

78/8 Sand and gravel resources of the Highland Region.
W. Mykura, D. L. Ross and F. May.
ISBN 0 11 884050 9 £3.00

Dd 696467 K8

Typeset for the Institute of Geological Sciences by Willsons Printers (Leicester) Limited

Printed in England for Her Majesty's Stationery Office by Commercial Colour Press, London E7